Nondestructive Evaluation of Materials by Infrared Thermography

Xavier P. V. Maldague

Nondestructive Evaluation of Materials by Infrared Thermography

With 138 figures

Springer-Verlag
London Berlin Heidelberg New York
Paris Tokyo Hong Kong
Barcelona Budapest

Professor Xavier P. V. Maldague, PhD
Université Laval, Québec, Canada

Cover illustration: TNDE configuration for the inspection of large components mounted on a moving slide.

ISBN-13:978-1-4471-1997-5   e-ISBN-13:978-1-4471-1995-1
DOI: 10.1007/978-1-4471-1995-1

British Library Cataloguing in Publication Data
Maldague, Xavier
  Non-destructive Evaluation of Materials
  by Infrared Thermography
  I. Title
  620.1127
ISBN-13:978-1-4471-1997-5

Library of Congress Cataloging-in-Publication Data
Maldague, Xavier, 1959–
  Non destructive evaluation of materials by infrared thermography/
by Xavier Maldague.
      p.   cm.
  Includes bibliographical references and index.
  ISBN-13:978-1-4471-1997-5 (U.S.) : $134.40 (est.)
  1. Engineering inspection—Automation.   2. Thermography-
Industrial applications.   3. Non-destructive testing—Industrial
applications.   4. Infrared technology—Industrial applications.
I. Title.
TS156.2.M35   1992   92-33616
620.1′127—dc20   CIP

Typeset by Thomson Press (India) Ltd., New Delhi
69/3830-543210 Printed on acid-free paper

quand le midi de ses feux bienfaisants
Ranime par degrés mes membres languissants,
Il me semble qu'un Dieu dans tes rayons de flamme,
En échauffant mon sein, pénètre dans mon âme!

Alphonse de Lamartine
*Méditations **29***

To
my Family,

to
Marthe, Lucas

Contents

Introduction

With national trade barriers falling, causing the expansion of the competitive global market, the question of *quality control* has become an essential issue for the 1990s. The time where the promise was to replace a product if it does not work seems to have passed; what is more important now is not so much a reduction in what is going wrong but an increase of what is going right the first time (Feigenbaum 1990). This new trend is sometimes referred to as *total quality*.

Among the many advantages of this zero-defect manufacturing policy, we can enumerate (Laurin 1990): superior marketability of wholly dependable products, enormous gain in productivity, elimination of wasteful cost in replacing poor quality work and retrofitting rejected products from the field. Although total quality is a relatively new and attractive concept for mass products such as cars, consumer electronics and personal computers, in many fields, mainly aerospace and military, it has been the rule for years because of security reasons.

One of the major efforts to reach this quality concept is to implement inspection tasks along the production line through *computer vision*. Unlike manual inspection which is a variable process prone to fatigue and lack of motivation, automatic vision can be deployed in hostile environments and allows a uniform and repeatable judgement. The increase in quality resulting from careful inspection leads to a reduction in the number of defective parts passed to subsequent operations. The quality is hence directly incorporated into the product (Wallace 1988). Since fewer unreliable and defective products are delivered, the customer's satisfaction grows and, following the snowball effect, the market share increases.

This context of quality control of industrial processes and automatic inspection, which continues to move further from traditional methods based on personal expertise acquired after many years of practice (Boogaard 1992) forms the basis of this book. More specifically, the scope of this book is infrared thermography inspection, and the many advantages in its application to nondestructive evaluation (NDE).

Nondestructive evaluation (NDE), nondestructive testing (NDT) and nondestructive inspection (NDI) are names to describe methods for testing without damage. In fact these terms all mean much the same and attempts to assign different meanings to each has never been successful, for example (Scott 1990): NDT (discovery of defects, practical aspects of the techniques); NDI (quantification of defects); NDE (assessment of the importance of defects). In this book NDE will be used since it is an acronym often found in the research field from which much of the work presented here comes. For instance, research institutions generally use the name NDE (e.g. *Journal of NDE, Research in NDE, Review of progress in quantitative NDE*, etc.).

Unlike visible spectra images (wavelength spectrum: $0.35–0.75\,\mu m$) which are produced by reflection and reflectivity differences, infrared images (wavelength spectrum: $0.75–100\,\mu m$, if taken in a broad sense) are produced by a self-emission phenomenon and also by variations in emissivity. Consequently, inspection processes using infrared techniques will be different from traditional processes based on interpretation and analysis of visible spectrum images.

The military were the first to take an interest in applications of the infrared (see e.g. Stillwell 1981; Girard and Algazi 1985; Strickland and Gerber 1986; Fredal et al. 1987; Richardson and Schafer 1987; Herby 1988; Mao and Strickland 1988), mainly to produce vision systems able to reveal the presence of potential targets in poor visibility conditions (e.g. at night or in fog).

The potential of nondestructive evaluation of materials by infrared thermography (or TNDE) is being increasingly exploited, especially since the availability, in the 1980s (Reynolds 1988), of commercial infrared cameras whose video signals are compatible with black-and-white television standards (the RS-170 standard in particular). Nevertheless, a lot of research on infrared image processing and inspection methods is still being undertaken. In fact, the complete commercial infrared systems that can now be purchased are still relatively limited in their various functions.

With this book we want to present the reader with an up-to-date coverage of infrared thermography applied in the context of industrial productivity and quality through automated inspection and control. Particular emphasis is placed on three fundamental aspects of TNDE: thermal methods, image processing and quantitative characterization. The TNDE procedure can be thought of in two stages. First, thermal methods are deployed to perform thermal inspection, and second, defect(s) detection and characterisation operations proceed. It may be noted that for the majority of industrial applications, the detection step is sufficient (Barre 1988).

Of relevance, for some applications, is the aspect of quantitative characterization which allows one to understand what is happening inside the inspected component. Proper modelling based on the work of many researchers (e.g. Parker et al. 1961; Vavilov 1980; Williams et al. 1980; Cielo 1984; Reynolds and Wells 1984; Sayers 1984; Monti and Mannara 1985; Degiovanni 1986; Hsieh and Kassab 1986; Balageas et al. 1987a; Dartois et al. 1987; Balageas 1991; Krapez 1991a; David et al. 1992) is introduced, always keeping in mind the book is intended for a larger audience not necessarily familiar with complex mathematics. Following this point, the essentials for comprehension are given, while references will provide additional information (see for instance Burleigh (1987,

1988a, b) for an extended bibliographical review, although the reference section at the end of the book is already quite complete). This type of modelling permits one to select appropriate thermographic inspection parameters.

In addition to quantitative characterization, the main *classical* methods needed to detect defects using infrared images are described, including several dedicated image processing techniques needed either to detect defects, enhance their visibility or reduce infrared image noise. However, the TNDE field is still relatively young and it should be noted that the processing methods are often simply adapted from classical video image interpretation.

The intended audience is threefold. First the book is intended for students who want to learn in depth about TNDE. In this respect this publication can complement an advanced material engineering curriculum. Due to the complete coverage of practical aspects, laboratory sessions can be envisaged as well. Industrial engineers will find valuable information in order to consider TNDE as an alternative inspection tool or for actual deployment on the plant floor. Finally, researchers will appreciate extended coverage of the subject in one volume with numerous references to relevant papers. Some readers may find surprising the many departures from fundamental aspects to more practical concerns. This is intentional, as the application of science does need experimental expertise. This is one aspect this book wants to satisfy.

We start, in Chap. 1, with an introduction to the field of infrared thermography, followed by a presentation of important theoretical points of radiometry and heat transfer modelling (Chap. 2). Experimental aspects concerning the thermographic inspection station are considered in Chap. 3. Experimental methods, thermal image (thermogram) processing for inspection of internal material structure and detection of subsurface defects are covered in Chaps 4 and 5. Quantitative characterization is the subject of Chap. 6. Inspection of low emissivity planar objects is described in Chap. 7. In Chap. 8, some infrared imaging methods for thermal diffusivity measurement are introduced, while the new concepts of thermal tomography are reviewed in Chap. 9. Chapter 10 is dedicated to some considerations for the TNDE inspection of nonplanar surfaces. Examples of high temperature applications of infrared thermography are described in Chap. 11.

This book would not have been possible without the cooperation of many individuals. In this respect I am grateful to Dr. Paolo Cielo of IMI who introduced me to the field of TNDE inspection while I was working there a few years ago. The IMI is the Industrial Materials Institute located in Montréal, Québec a division of the National Research Council of Canada. I would like also to express my gratitude to Professor Denis Poussart of Université Laval who encouraged me to go ahead in this publication adventure. Special thanks to Mr. Gaston Guay of our E.E. Department who prepared most of the drawings for the figures. I wish also to thank the postgraduate students of the Laboratoire de Vision et Systèmes Numériques (LVSN) who helped in preparing some of the material needed for the figures. I also offer my appreciation to Professor

Vladimir Vavilov of the Tomsk Polytechnic Institute (CIS, former Soviet Union) who collaborated in the writing of Chap. IX during his sabbatical leave to the LVSN.

Support of our E.E. Department is acknowledged as well as the financial support of the NSERC (National Sciences and Engineering Research Council of Canada), the FCAR (Fonds pour la Formation des Chercheurs et l'Aide à la Recherche of the Québec province) and the Centre of Excellence IRIS (Institute for Robotics and Intelligent Systems); this support was greatly appreciated. Finally, the invaluable support from my family, especially my wife Marthe, was essential to the success of this work.

Xavier Maldague,
Québec, March 1992

Figure Credits

Fig. 1.2   Adapted from Cielo et al. (1987a), Fig. 6, p 456.
Fig. 1.4   Adapted from Vavilov and Taylor (1982), Fig. 1, p 248.
Fig. 1.5   Adapted from Vavilov and Taylor (1982), Fig. 2, p 249.
Fig. 1.6   Adapted from Tretout (1987), Figs 2 and 3, p 48.
Fig. 1.7   Adapted from Loubet (1987), Figs 1–3.
Fig. 1.9   Adapted from Gaussorgues (1984b), p 308.
Fig. 2.1   Adapted from Nicodemus (1967), Fig. 9, p 288.
Fig. 2.2   Reproduced with permission from Maldague and Dufour (1989), Fig. 7, p 876.
Fig. 2.4   Adapted from Tossell (1987), Fig. 1, p 102.
Fig. 3.2   Adapted from AGEMA (1984), Fig. 2.16.
Fig. 3.3   Adapted from Holmsten (1986), Fig. 14, p 86.
Fig. 3.21   Adapted from Gaussorgues (1984b), p 283.
Fig. 3.23   Adapted from Maldague et al. (1991b), Fig. 2, p 23.
Fig. 3.25   Reproduced with permission from Maldague et al. (1991b) Fig. 4, p 24.
Fig. 3.26   Reproduced with permission from Maldague et al. (1991b), Fig. 5, p 25.
Fig. 4.1   Reproduced with permission from Cielo et al. (1987a), Fig. 1, p 453.
Fig. 4.3   Adapted from Cielo et al. (1987b), Fig. 7, p 740.
Fig. 4.5   Reproduced with permission from Krapez et al. (1987), Fig. 2, p 397.
Fig. 4.6   Reproduced with permission from Cielo et al. (1987a), Fig. 2, p 454.
Fig. 4.7   Reproduced with permission from Cielo et al. (1987a), Fig. 3, p 454.
Fig. 4.8   Reproduced with permission from Cielo et al. (1987a), Fig. 4, p 455.
Fig. 4.9   Reproduced with permission from Cielo et al. (1987a), Fig. 5, p 455.
Fig. 4.10   Adapted from Cielo et al. (1988b), Fig. 1, p 405.
Fig. 4.11   Adapted from Maldague et al. (1987b), Fig. 1, p 46.
Fig. 4.12   Adapted from Maldague et al. (1987b), Fig. 2, p 47.
Fig. 4.15   Adapted from Cielo et al. (1987a), Fig. 9, p 457.
Fig. 4.27   Adapted from Maldague et al. (1990b), Fig. 1, p 723.

Fig. 4.28   Adapted from Maldague et al. (1990b), Fig. 2, p 724.
Fig. 5.1   Reproduced with permission from Maldague et al. (1989b),
             Fig. 2, p 565.
Fig. 5.6   Adapted from Maldague et al. (1990c), Fig. 4, p 145.
Fig. 5.7   Reproduced with permission from Maldague et al. (1990a),
             Fig. 4, p 166.
Fig. 5.8   Reproduced with permission from Maldague et al. (1990c),
             Fig. 5, p 146.
Fig. 5.9   Adapted from Maldague et al. (1990c), Fig. 6, p 147.
Fig. 5.10   Reproduced with permission from Maldague et al. (1990a),
             Fig. 5, p 167.
Fig. 5.11   Adapted from Maldague et al. (1990c), Fig. 7, p 147.
Fig. 5.12   Reproduced with permission from Maldague et al. (1990a),
             Fig. 6, p 169.
Fig. 5.13   Reproduced with permission from Maldague et al. (1990c),
             Fig. 9, p 150.
Fig. 5.14   Reproduced with permission from Maldague et al. (1990c),
             Fig. 10, p 151.
Fig. 6.5   Adapted from Krapez et al. (1991), Figs 2 and 3, p 89.
Fig. 6.6   Adapted from Balageas et al. (1987c), Fig. 4, p 463.
Fig. 6.7   Adapted from Krapez et al. (1991), Fig. 7, p 110.
Fig. 7.2   Adapted from Maldague et al. (1991a), Fig. 1, p 118.
Fig. 7.3   Adapted from Maldague et al. (1991a), Fig. 2, p 119.
Fig. 7.4   Adapted from Maldague et al. (1991a), Fig. 3, p 119.
Fig. 7.5   Adapted from Maldague et al. (1991a), Fig. 4, p 120.
Fig. 7.6   Adapted from Maldague et al. (1991a), Fig. 7, p 123.
Fig. 7.7   Reproduced with permission from Maldague et al. (1991a),
             Fig. 8, p 124.
Fig. 7.8a   Reproduced with permission from Maldague et al. (1991a),
             Fig. 9, p 124.
Fig. 8.1   Adapted from Delpech et al. (1990), Fig. 2, p 1.
Fig. 8.2   Adapted from Delpech et al. (1990), Fig. 3, p 2.
Fig. 8.3   Adapted from Delpech et al. (1990), Fig. 4, p 2.
Fig. 8.6   Adapted from Heath and Winfree (1989), Fig. 2, p 1614.
Fig. 8.7   Adapted from Heath and Winfree (1989), Fig. 3, p 1614.
Fig. 8.8   Adapted from Heath and Winfree (1989), Fig. 5, p 1616.
Fig. 9.3   Adapted from Vavilov et al. (1992), Fig. 5.
Fig. 10.1   Reproduced with permission from Maldague et al. (1991c),
             Fig. 1, p 241.
Fig. 11.3a   Reproduced with permission from Maldague et al. (1987a),
             Fig. 4-a, p 123.
Fig. 11.4   Adapted from Maldague et al. (1987a), Fig. 5, p 124.
Fig. 11.5c   Adapted from Maldague et al. (1987a), Fig. 6, p 124.
Fig. 11.6   Reproduced with permission from Maldague et al. (1987a),
             Fig. 7, p 125.

COLOUR SECTION

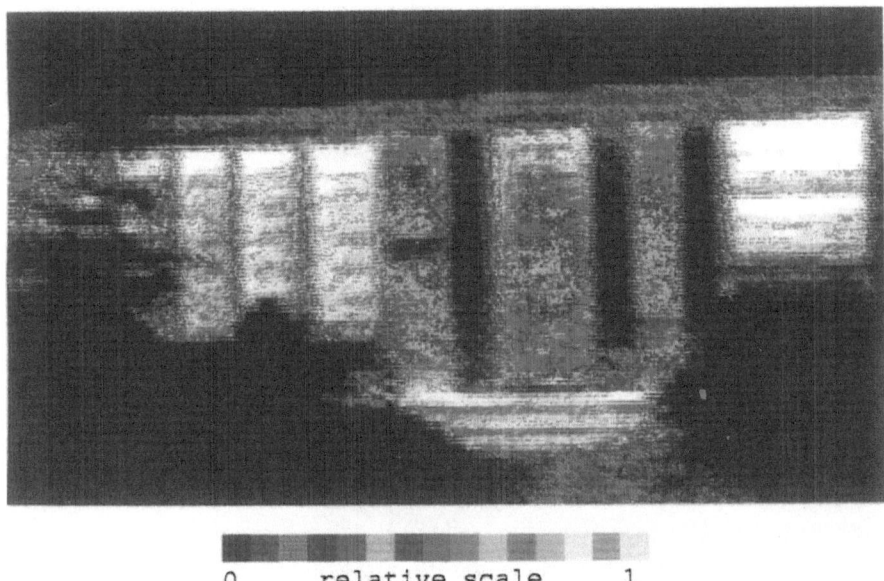

0   relative scale   1

Fig. 1.1. Passive TNDE: investigation of home thermal insulation efficiency. Severe heat losses through the front door are clearly visible on this thermogram. Notice spurious reflections on the icy path in front of the stairs. Exact temperature can be computed only if emissivities are known (cf. Sect. 2.1).

plate with grooves

25.0C   26.0C   27.0C

Fig. 3.18. Spatial geometry effects: image of a plate with grooves (1 cm apart, 0.3 mm width, 0.1 mm depth). The plate is uniformly heated in a warm water bath whose temperature is close to infrared camera temperature in order to limit radiometric distortion effects.

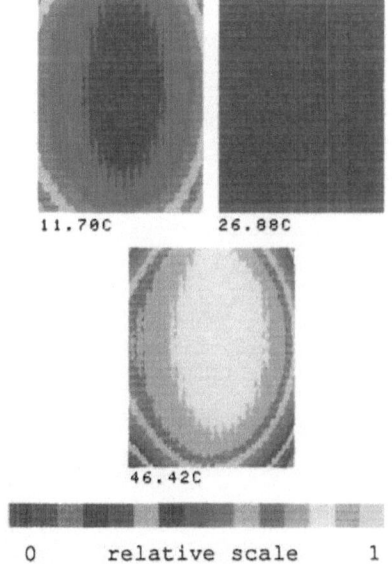

11.70C   26.88C

46.42C

0   relative scale   1

Fig. 3.24. Effects of radiometric distortions: images recorded on a thick brass plate of high thermal conductivity heated at uniform temperature by immersion in a water tank (11.70 °C, 26.88 °C, 46.42 °C).

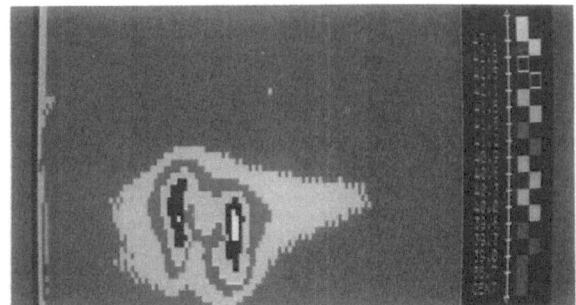

Fig. 4.1. Impact damage on eight-ply graphite epoxy plates. The field of view is 5.5×5.0 cm; impact energy was about 6J. Orientations: **a** $(0, \pm 45, 90_s$; **b** $(0,90)_{2s}$.

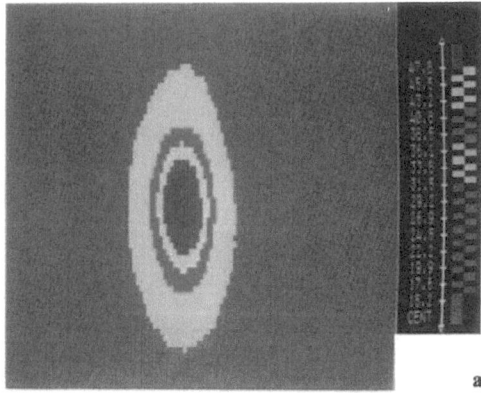

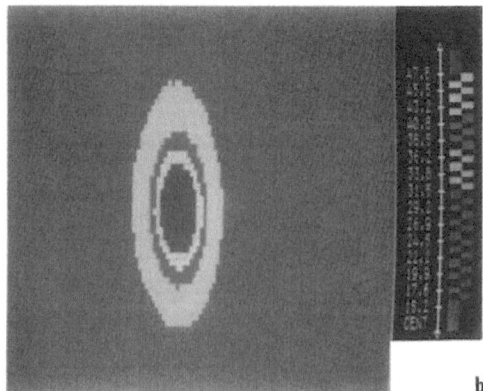

Fig. 4.3. Elliptical thermal patterns obtained with point laser heating on a graphite epoxy plate cured with pressure **a** normal and **b** low. Image **c** shows the difference between the two patterns. The field of view is $25\,cm^2$.

Fig. 4.6. Raw thermal images recorded on a graphite epoxy plate containing Teflon™ implants. Top row: implant size **a** 20 × 20 mm, **b** 10 × 10 mm **c** 3 × 3 mm, inserted 3 mm below the front surface. Bottom row: Implant size 20 × 20 mm, inserted **d** 0.3 mm, **e** 1.12 mm, **f** 2.25 mm below the front surface.

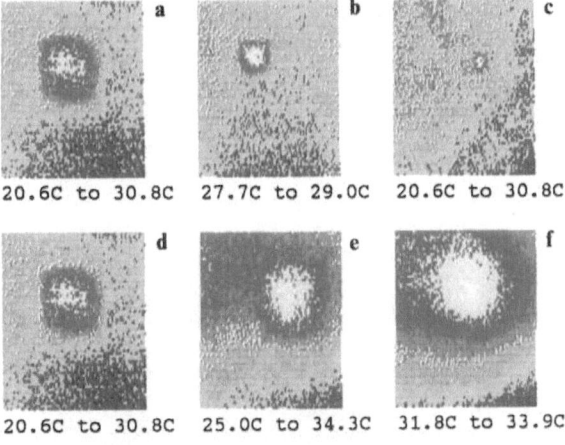

20.6C to 30.8C   27.7C to 29.0C   20.6C to 30.8C

20.6C to 30.8C   25.0C to 34.3C   31.8C to 33.9C

Fig. 4.7. Example of image processing making use of the spatial reference technique. Image **b** obtained over a sound area is subtracted from image **a** recorded over a defect of 20 × 20 mm, 2.25 mm below the surface (defect in Fig. 4.6f) thus improving defect visibility in **c**.

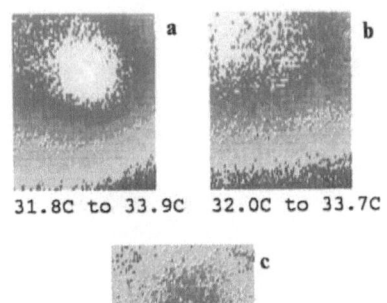

31.8C to 33.9C   32.0C to 33.7C

-0.5C to +0.7C

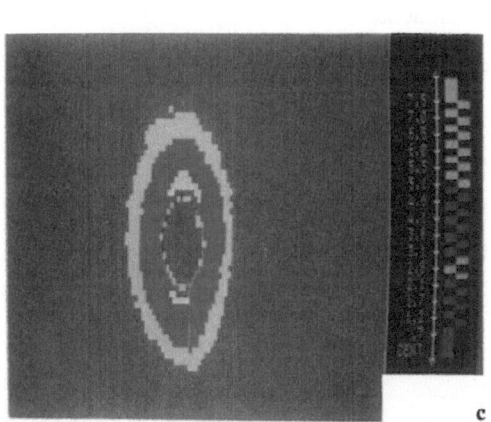

c

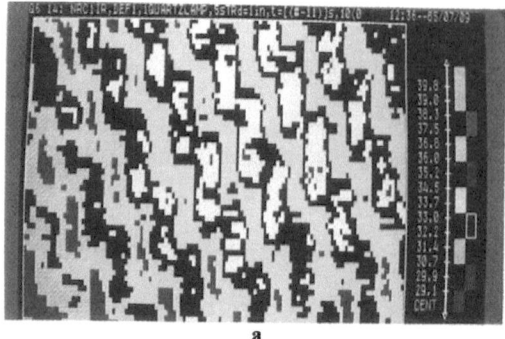

a

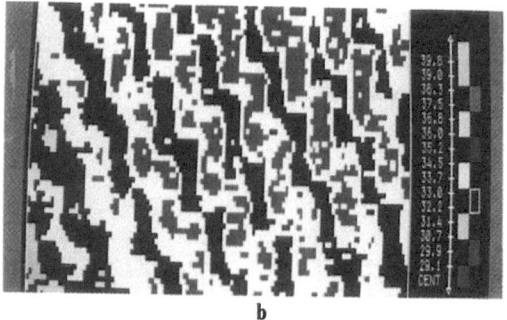

b

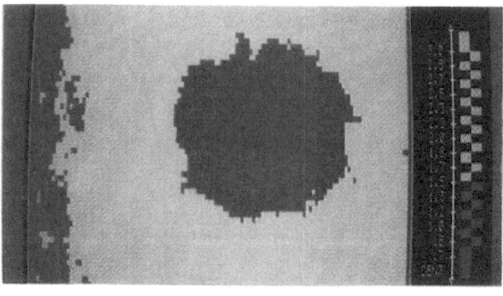

c

Fig. 4.8. Example of image processing making use of the temporal reference technique. Images **a** and **b** were recorded over the same zone containing a defect of 10×10 mm, 0.5 mm under the front surface in extreme conditions of thermal noise (parasitic reflections) 3 s and 5 s respectively after beginning of heating. The subtracted image **c** offers an enhanced defect visibility.

triangular defect

25.5C	29.4C	33.2C

Fig. 4.12. Detection of lack of adhesive (zone of triangular shape) in an Al–foam panel. Effect of the poor heating uniformity is apparent on this thermogram.

defect

sound area

subtraction

28.7C	32.4C	36.0C

Fig. 4.13. Spatial reference technique used to improve defect visibility. Top, raw image recorded over the defect. Middle, image recorded over a sound area. Bottom, the subtracted image delineates clearly the position of the triangularly-shaped zone with missing bonding.

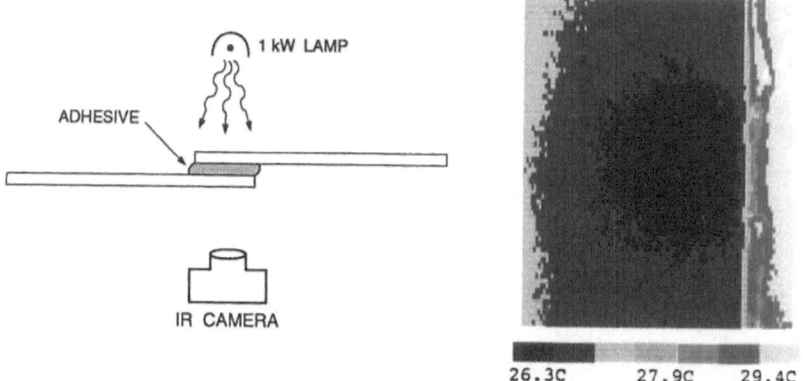

Fig. 4.14. Thermographic inspection in transmission for an Al laminate (two sheets bonded). Left, experimental apparatus; right, thermal image showing a zone without adhesive of 1.5 × 1.5 cm. The field of view is 4 × 5 cm.

Fig. 4.15. Thermographic inspection by propagation of a cold front for the inspection of bonded Al structures. Left, experimental apparatus; right, thermal image showing a zone of 1.5 × 1.5 cm without adhesive.

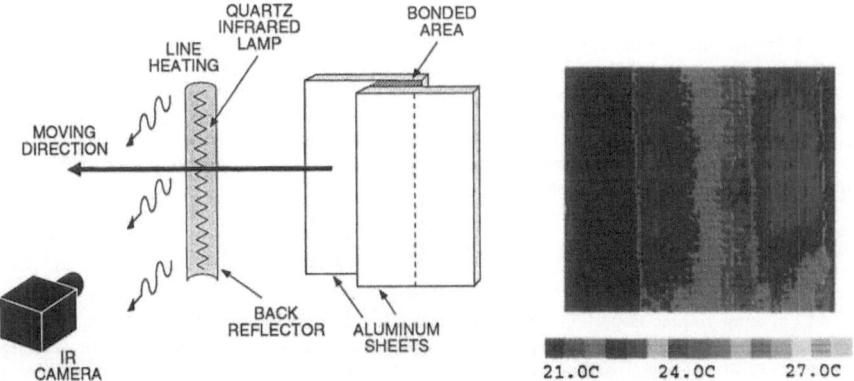

Fig. 4.16. Thermographic inspection of a bonded line between two aluminium sheets using the mobile configuration. Left, experimental apparatus; right, the thermogram shows a bonding defect at the image centre. The lack of uniformity at the bottom of the thermogram is due to insufficient heating in this area (the lamp used – 530 W, 18 cm long – was a little short with regard to the specimen size).

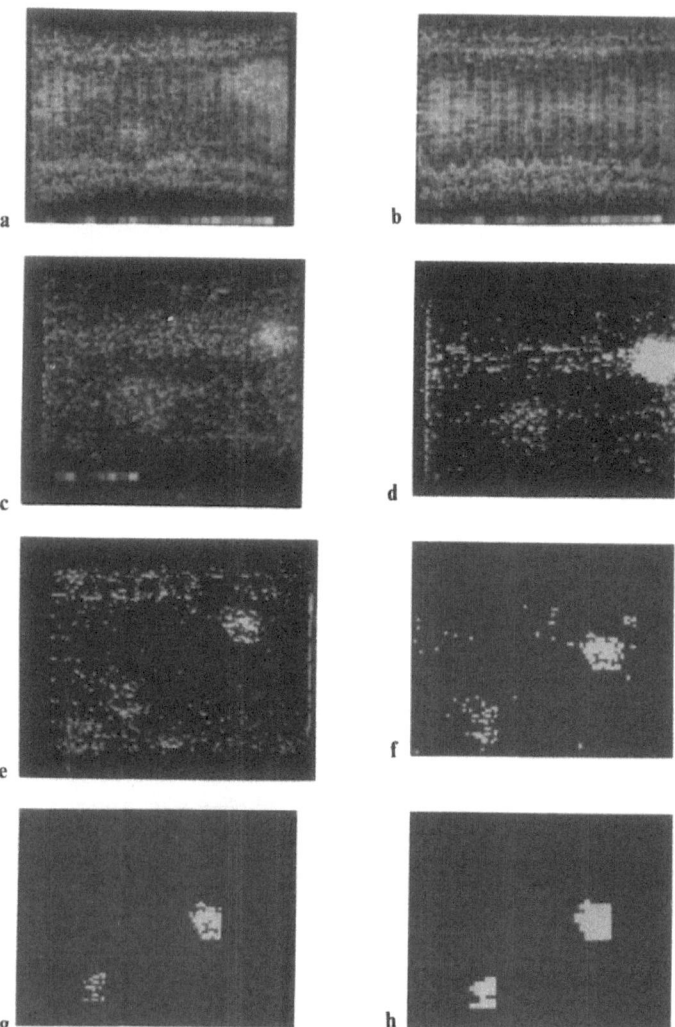

Fig. 4.19. Steps needed for defect extraction from images recorded over an aluminium laminate (bonding defect). Raw image recorded at $t = 3.04$ s after beginning of heating.
a above a defect. **b** above a sound area. **c** ($= \mathbf{a} - \mathbf{b}$) two defects are visible despite the presence of strong noise. **d** binary image of **c**. **e** binary image recorded at $t = 3.33$ s. The same steps as for d have been applied on this image. **f** is obtained from **d** and **e**, gross localization of two defects thanks to their motion in the field of view (see the text). **g** noise suppression through erosion. **h** defect reconstruction through dilation.
Note a 530 W, 18 cm long lamp is used for this test.

Opposite page:

Fig. 4.17. Thermographic inspection using a line of air jets. Top, experimental apparatus; bottom, the thermogram shows on left an Al–foam panel without defect and on right a similar panel with a circular 4 cm diameter lack of adhesive defect visible at the image centre.

Fig. 4.18. Mobile configuration, Al–foam panel, line heating 30 cm long, 2500 W. Top, experimental apparatus; bottom, a few thermograms recorded at specific times are displayed while the specimen was moving across the field of view. Careful examination of the sequence reveals three defects. Same specimen as Fig. 4.30b where three unbonded areas are visible.

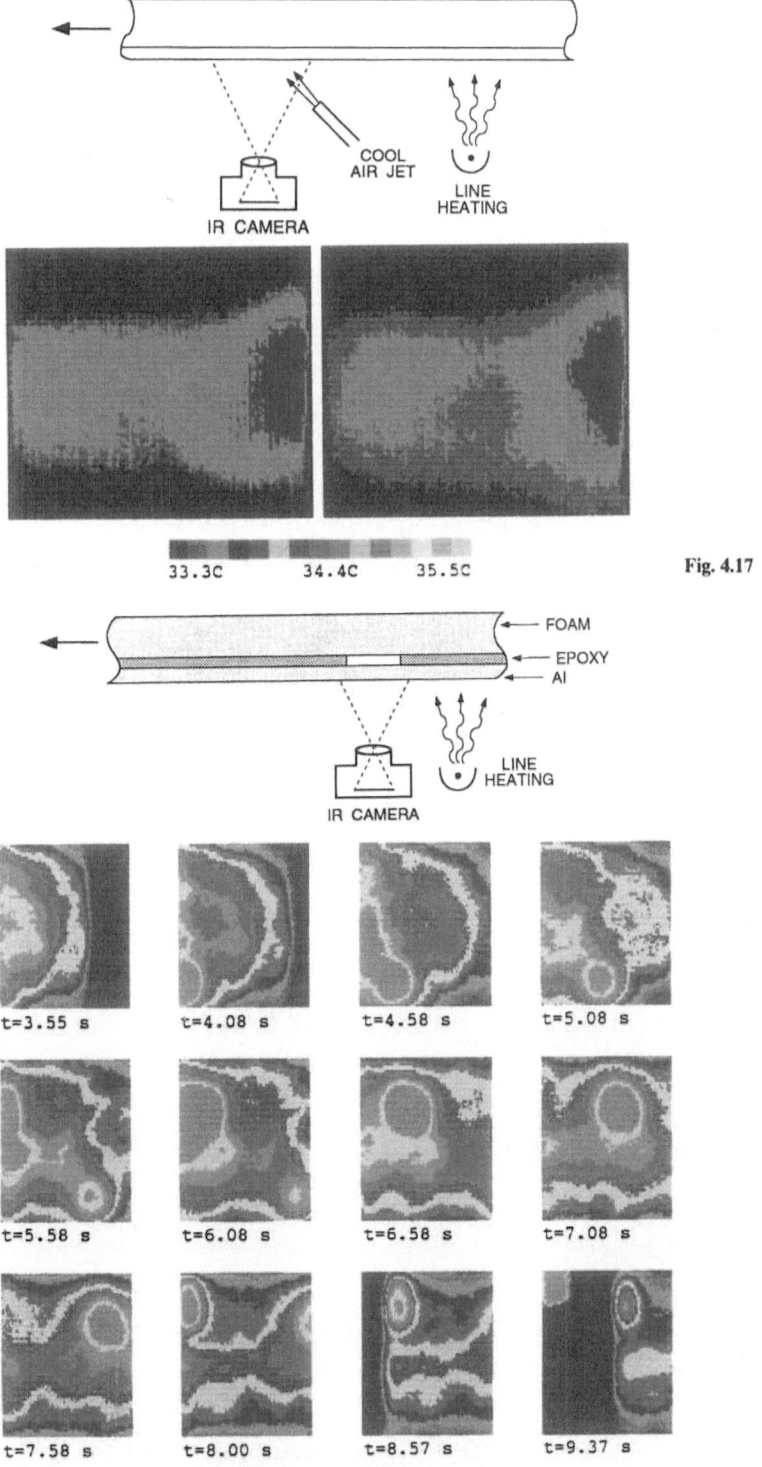

COOL
AIR JET

LINE
HEATING

IR CAMERA

33.3C   34.4C   35.5C

Fig. 4.17

FOAM
EPOXY
Al

LINE
HEATING

IR CAMERA

t=3.55 s   t=4.08 s   t=4.58 s   t=5.08 s

t=5.58 s   t=6.08 s   t=6.58 s   t=7.08 s

t=7.58 s   t=8.00 s   t=8.57 s   t=9.37 s

22.0C   26.0C   30.0C

Fig. 4.18

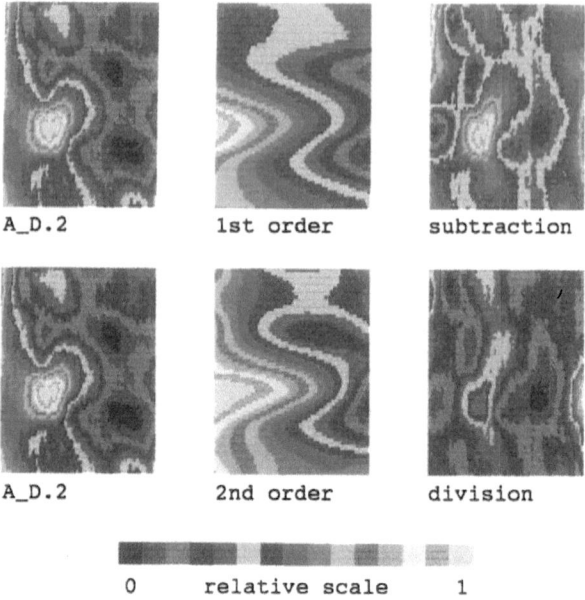

A_D.2   1st order   subtraction

A_D.2   2nd order   division

0   relative scale   1

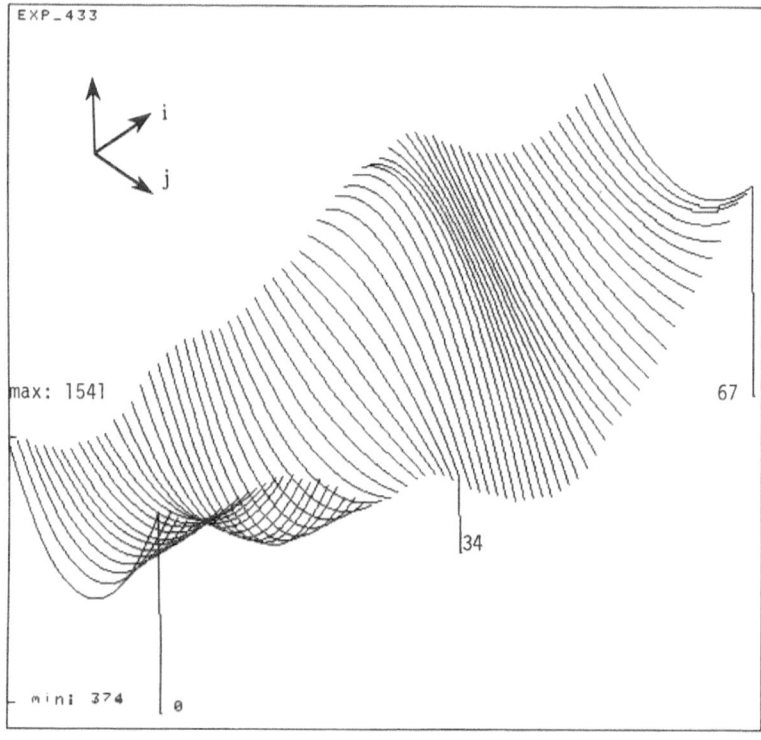

Fig. 4.20. Test of trend removal procedure. Top, series of images with image A_D.2 being the starting image; middle and right pairings are first and second order synthetic images; both a subtraction and division images are shown. Bottom, three-dimensional plot of the second order synthetic image.

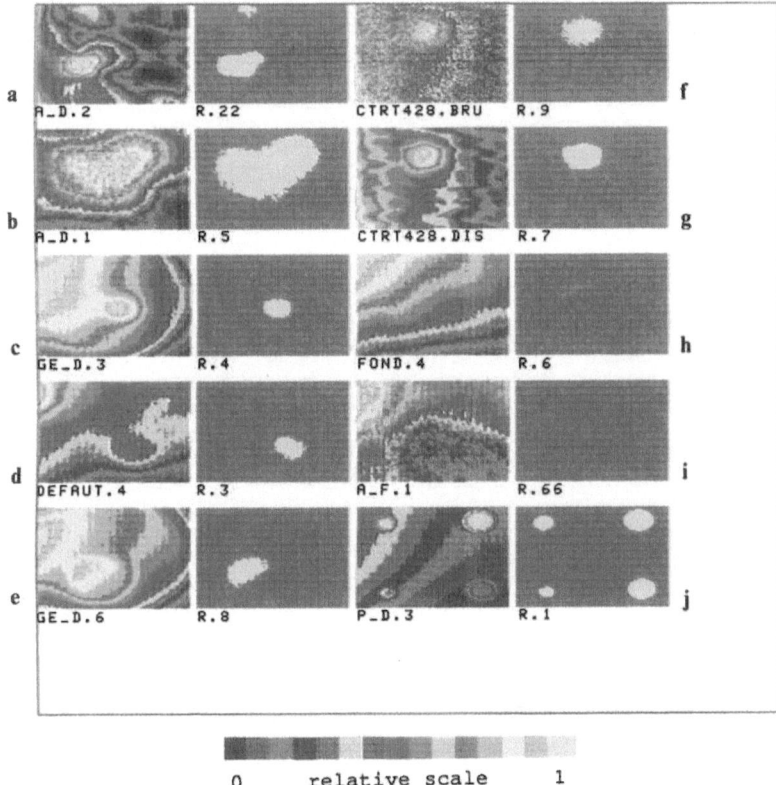

a R_D.2   R.22   CTRT428.BRU   R.9   f

b R_D.1   R.5   CTRT428.DIS   R.7   g

c GE_D.3   R.4   FOND.4   R.6   h

d DEFAUT.4 R.3   A_F.1   R.66   i

e GE_D.6   R.8   P_D.3   R.1   j

0   relative scale   1

Fig. 4.29. Segmentation results, static configuration. MND = 23. The field of view is 5 × 5 cm for all cases except **j** where it is 12 × 12cm. **a,b** Aluminium honeycomb (aluminium sheet 0.5 mm thick, epoxy bonded on a honeycomb core). A bonding defect (lack of epoxy) is present at the bonding interface. Two images having different field of view are shown with the detected defect having, of course, the same shape. A false reading is detected in **a**, at the top of the image. **c,d,e** Graphite epoxy panel with artificial defects (Teflon™ implant, 10 mm in diameter) inserted at different depths under the surface: **(c,d)** 1 mm, **(e)** 0.5 mm. Note the strong radiometric distortions on raw images **(d,e)**. **f,g** Graphite epoxy panel as in **c,d**. These images are obtained differently, they are in fact thermal contrast images which serve for the quantitative analysis that will be studied in Chap. 6. **f** is the raw image and **g** is the smoothed image (using the technique presented in Chap. 3). This demonstrates the good noise immunity of the algorithm. **h,i** Graphite epoxy panel without defect. **j** Plexiglass™ plate in which holes were drilled at different depths (Buchanan et al. 1990) from the back surface (1.12 mm top row and 2.25 mm bottom row, with diameters 10 mm on left and 20 mm on right). These are raw images where the effects of the apparent spatial resolution are not corrected, thus explaining the elliptical rather than circular shapes observed. These defects are truly artificial; this plate is tested because it has many known defects.

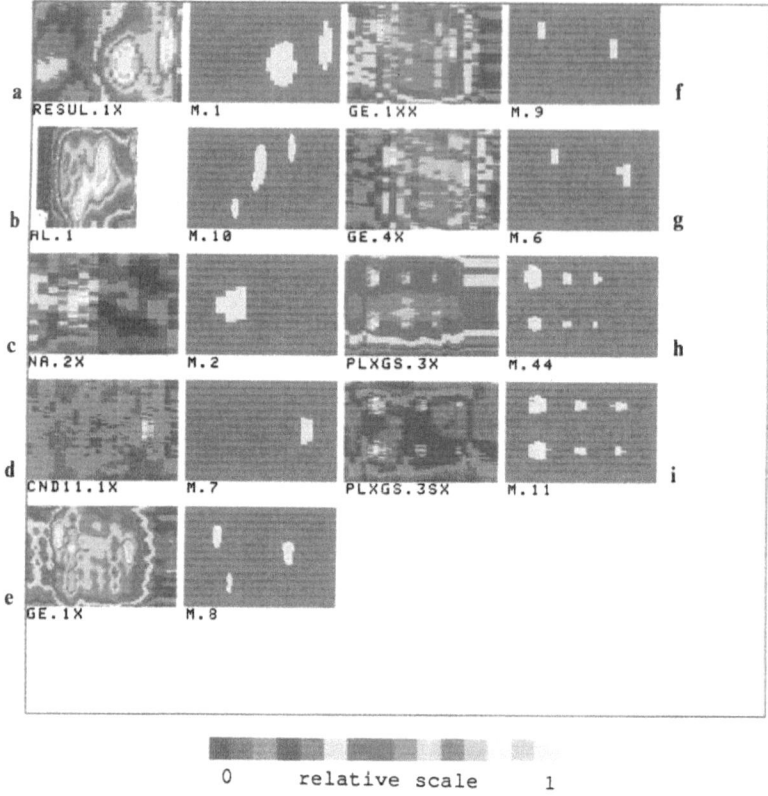

a RESUL.1X   M.1   GE.1XX   M.9   f

b AL.1   M.10   GE.4X   M.6   g

c NA.2X   M.2   PLXGS.3X   M.44   h

d CND11.1X   M.7   PLXGS.3SX   M.11   i

e GE.1X   M.8

0   relative scale   1

Fig. 4.30. Segmentation results, mobile configuration, reconstructed images. **a,b** Aluminium laminate (aluminum sheet 0.5 mm thick, epoxy bonded on a foam core). Two bonding defects (lack of epoxy) are present at the interface. Field of view, 12 cm height on the complete panel length (40 cm). These panels are used, for instance, in the transport industry, in the construction of refrigerated vehicles. **c** Aluminium honeycomb panel (**a,b**), field of view: 12 cm height on the complete panel length (15 cm). **d** Graphite epoxy panel with artificial defect (Teflon™ implant, 10 mm in diameter inserted 0.5 mm under the surface). **e** Graphite epoxy panel with two artificial defects (Teflon™ implant, 10 mm in diameter, inserted 1 mm under the surface). A trend removal procedure was used (Sect. 4.4.1) before calling the segmentation algorithm with the goal of increasing the visibility of the defects. On the contrary, a false reading appears at the centre of the segmented image. Field of view: 12 cm height (complete panel length 17 cm). **f,g** Same panel as in **e** for two different runs (reconstructed images without previous trend removal procedure). **h,i** Plexiglass™ plate of Fig. 4.29j **h** Reconstructed image as is; **i** spatial subtraction technique studied previously: the image shown was obtained after subtraction from a reconstructed image of a Plexiglass™ plate without defect. As expected, **i** is "cleaner". Nevertheless the algorithm performed well in both cases and reveals the six drilled holes of 20, 10 and 5 mm diameter respectively from left to right, located at 1.12 mm beneath the surface (top row) and 2.25 mm beneath the surface (bottom row). Note that the smallest defect (5 mm diameter, 2.25 mm depth) is at the detectability limit since its radius/depth ratio is about 1.1 in this isotropic material. Lateral speed was different for **h** and **i**, explaining the different positions of detected defects in the images.

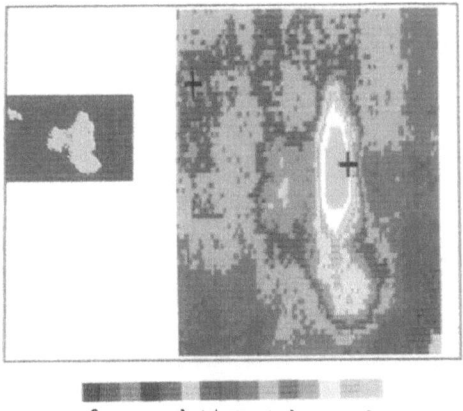

0 relative scale 1

Fig. 4.31. First part of the algorithm and segmented image of a graphite epoxy panel damaged by impacts. Crosses indicate the position of the defects (seeds) as located by the algorithm. What seems to be a false reading is visible on the left, although the delamination can stretch over large distances. Static configuration, field of view 4 × 4 cm.

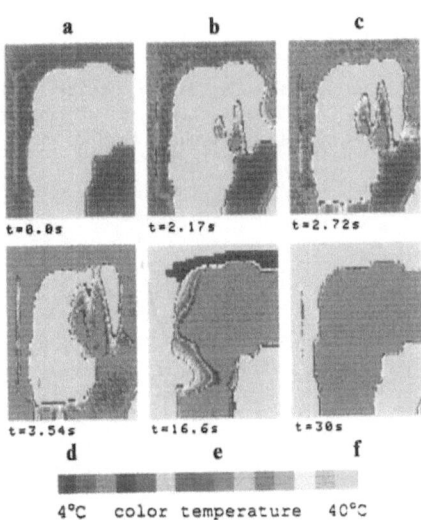

a   b   c

t=0.0s   t=2.17s   t=2.72s

d   e   f

t=3.54s   t=16.6s   t=30s

4°C   color temperature   40°C
            scale

Fig. 5.2. Infrared image sequence recorded during the active inspection of the corroded pipe bend section of Fig. 5.1 using the transmission approach. Temperature transition from **a** uniformly hot (40 °C) to **f** uniformly cold, (6 °C). Water drain on right.

4C   13C   22C

Fig. 5.3. Infrared image recorded at $t = 2.72$ s during the cold-to-warm transition; the same pattern as during the warm-to-cold transition (Fig. 5.2c) is observed.

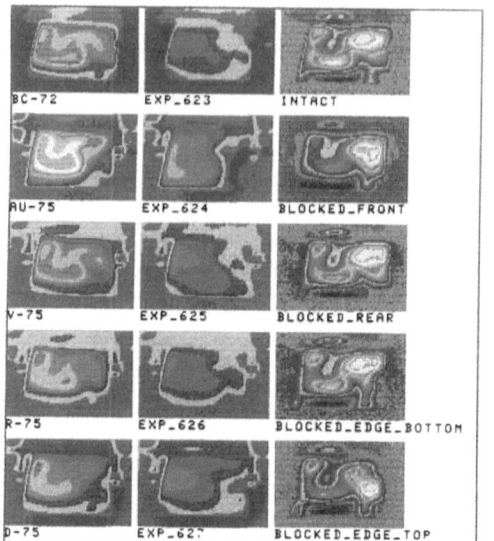

28.1C   29.7C   31.3C

Fig. 5.6. Thermal signatures obtained with the air rig for turbine blades intact and for some having blocked channels. Acquisition of two images, pressure side: one at $t = 1.8$ s after start of heating (left column) and one during the cooling of the blade (middle column, $t = 3.95$ s after start of cooling). The time subtraction image is shown for all cases (right column).

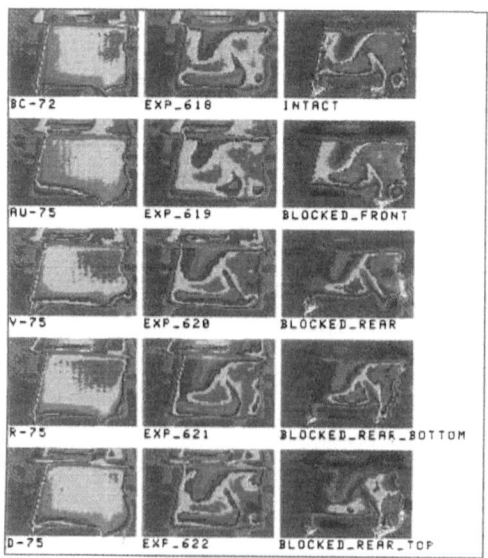

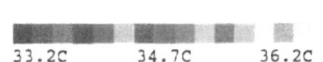

33.2C   34.7C   36.2C

Fig. 5.7. Thermal signatures obtained with the water rig for turbine blades intact and for some having blocked channels. Acquisition of two images, pressure side: one in the steady warm regime (left column) and one during the cooling of the blade (middle column, $t = 1.3$ s after starting cooling). The time subtraction image is shown for all cases (right column).

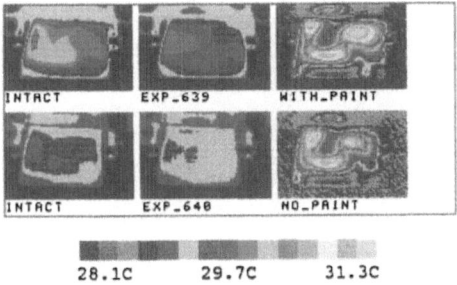

Fig. 5.9. Test on intact blades with the air rig. Upper row blades are black-painted to increase surface emissivity, bottom row blades are unpainted: the subtracted images on right reveal the same thermal signatures though attenuated in the unpainted case. Image positions and parameters are as in Fig. 5.6.

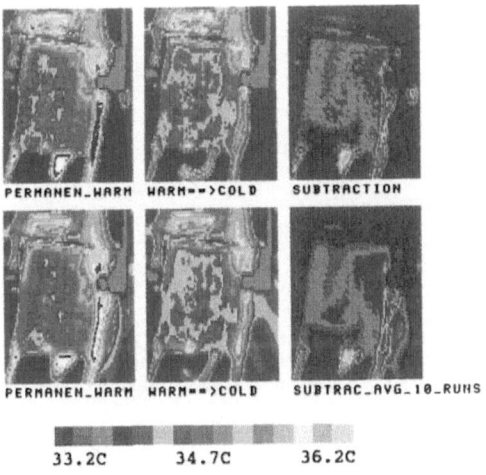

Fig. 5.10. Test on an intact unpainted blade with the water rig, acquisition of two images, pressure side. Top row: result of one run. Bottom row: average of ten consecutive runs. Image positions and parameters as in Fig. 5.7.

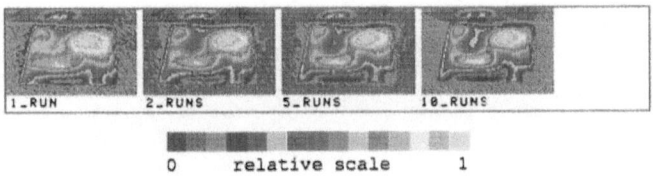

Fig. 5.11. Intact blade, air rig, same parameters as for Fig. 5.6. Increased signal-to-noise ratio is obtained by averaging many consecutive runs together.

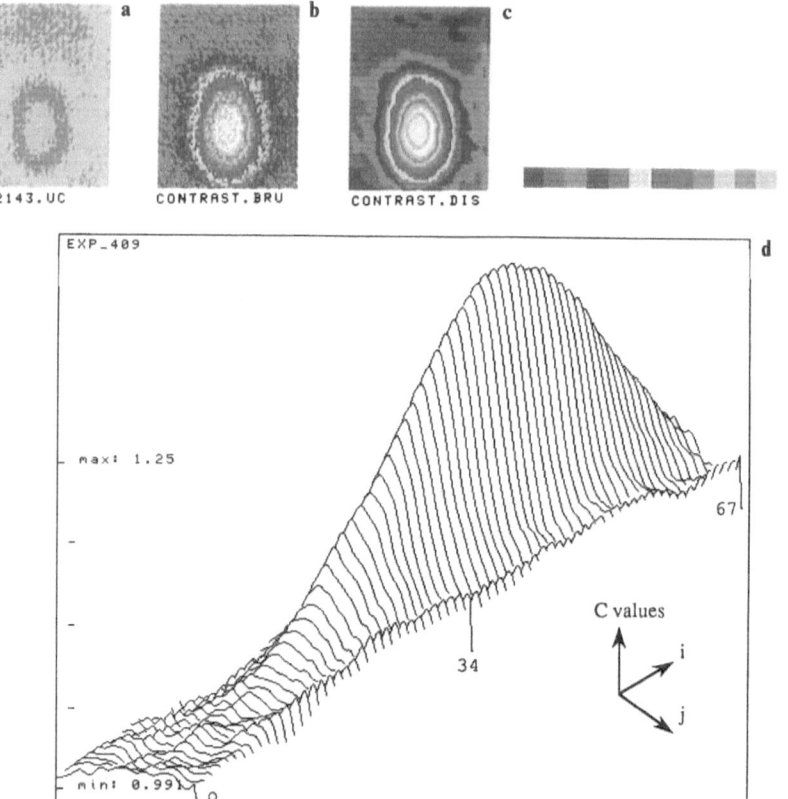

a
b
c

D_2143.UC   CONTRAST.BRU   CONTRAST.DIS

EXP_409

d

max: 1.25

67

C values

i

j

34

min: 0.99

O

Fig. 6.11. Example of Fig. 6.9: **a** raw images among those of time zone k_{max}; **b** contrast image C_{img} computed using Eq. (6.5); **c** image **b** smoothed using the technique of the Sect. 3.2.2; **d** three-dimensional plot of image **c**. The colour scale is relative (pixel intensity from 0 to 255).

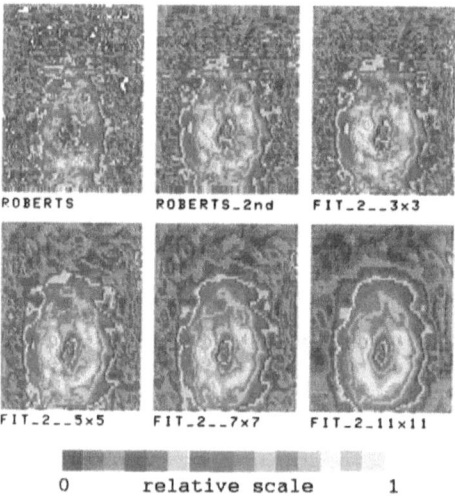

ROBERTS   ROBERTS_2nd   FIT_2__3×3

FIT_2__5×5   FIT_2__7×7   FIT_2_11×11

0   relative scale   1

Fig. 6.13. Gradient computations on the example of Fig. 6.11c,d.

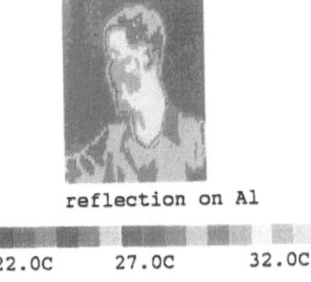

reflection on Al

22.0C   27.0C   32.0C

Fig. 7.1. Illustration of the problem of thermal reflections: the thermal image of the operator standing behind the infrared camera is reflected in the aluminium sheet which acts as a mirror. The camera is pointed at the sheet.

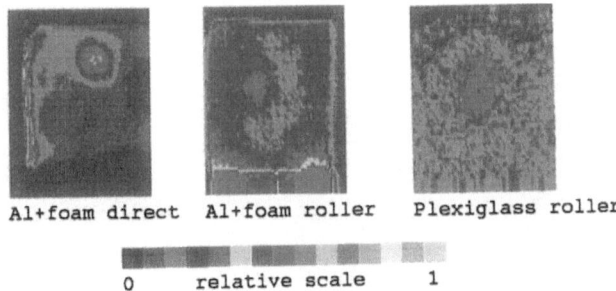

Al+foam direct   Al+foam roller   Plexiglass roller

0   relative scale   1

Fig. 7.8. Experimental results for the aluminium panel and Plexiglass plate (see text).

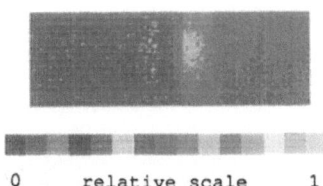

0   relative scale   1

Fig. 7.10. Reconstructed image of a defect (see text).

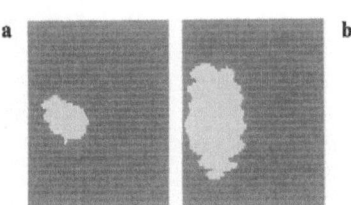

a   b

Fig. 7.11. a Segmented image of a defect obtained with thermal transfer imaging apparatus of Fig. 7.4 **b** Segmented image of the same defect obtained in direct observation with black-painted sample surface and no membrane. Same sample as for Fig. 7.9. Uncorrected size of field of view along rows and columns.

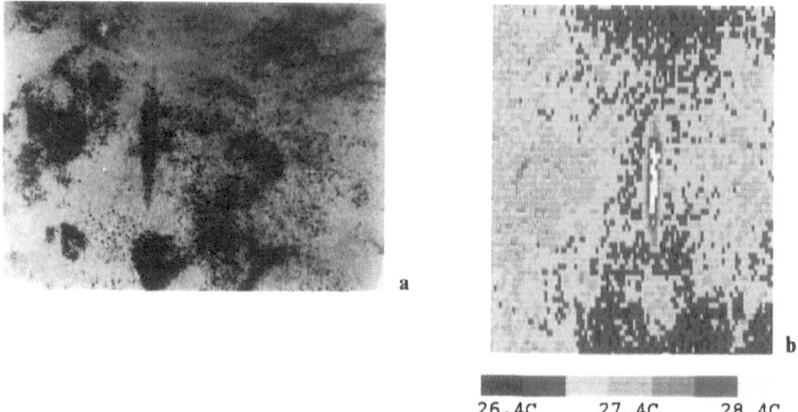

26.4C   27.4C   28.4C

Fig. 11.3. Comparison between **a** a video image of rolled-in-scale crust (visible image, diffuse illumination); **b** an emittance image recorded with the infrared camera.

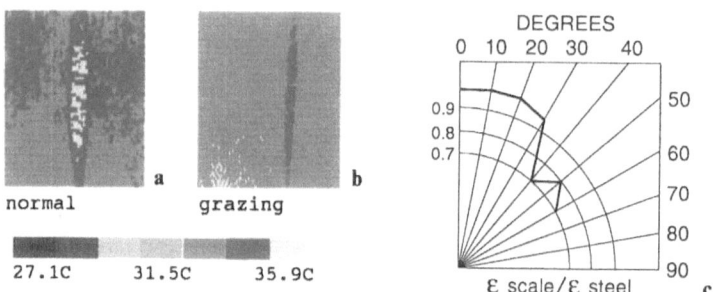

normal   grazing

27.1C   31.5C   35.9C

DEGREES
0 10 20 30   40

0.9
0.8
0.7

50
60
70
80
90

ε scale/ε steel   c

Fig. 11.5. Infrared image of the scale defect at **a** normal incidence and **b** near-grazing incidence. **c** Polar diagram showing the variation of the normalized emissivity ratio as a function of the observation angle with respect to the normal.

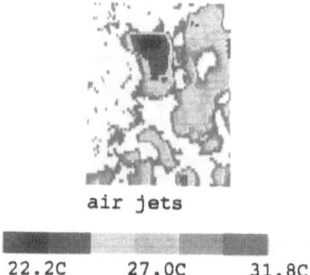

air jets

22.2C   27.0C   31.8C

Fig. 11.7. Cool thermal wave approach: an artificially oxidized, partially descaled plate is uniformly heated and subsequently surface cooled by an air jet. Colder regions correspond to the partially loose scale patches. Notice that the temperature scale is only indicative because of emissivity variations across the field of view.

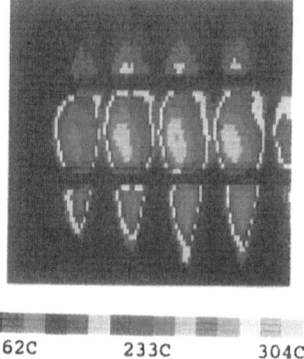

<162C   233C   304C

Fig. 11.9. Thermal image showing the internal structure of the heat exchanger unit of Fig. 11.8 during operation.

Chapter 1

Overview of Nondestructive Evaluation (NDE) Using Infrared Thermography

1.1 General Considerations

1.1.1 Short History of the Infrared

Even before Max Planck published his theory of radiation on 14 December, 1900, measurement of temperature was of concern to Mankind. The first law of thermo-dynamics introduces the energy conservation principle and explains that any (indus-trial) process consuming energy will see a great part of this energy be transformed into heat (following the law of entropy). Temperature is therefore an important para-meter to measure. The glass thermometer, invented by Galileo in 1593, was the first instrument for quantitative temperature measurement (Wise 1988). It allowed Herschel in 1800 to discover the infrared spectrum. Nowadays, temperature scale is ruled by international norms: the freezing point of gold (also known as the *gold point*) is the basic standard for the International Practical Temperature Scale (IPTS).

Sir William Herschel's history is worth a little elaboration (complete details can be found in Hudson (1969)). This great man was royal astronomer to King George III of England and accidentally discovered Uranus on 13 March, 1793. This was another "accident" which led to his discovery of infrared rays. At first he wanted to protect his eyes when observing the sun. For his experiment he used a prism which separated the various colours, from blue to red. Using a mercury thermometer, he noted that the maximum elevation of temperature occurred beyond the red band where no radiation was visible (Herschel 1800a, b). In fact this experiment had been done before but he was the first to notice that the distance where the heating is maximum is specifically located (i.e. depending of the wavelength). We know now this is related to Planck's (Eq. (2.2)) and Stefan's law:

$$\lambda_m = \frac{2898}{T}$$

This relation gives the peak wavelength λ_m (in μm) corresponding to the temperature T (in kelvin) for an emissive body. For example, observing a red-hot steel plate having a temperature of 1300 °C ($\simeq 1000$ K), the peak wavelength would be about 3 μm. In this case Herschel's thermometer would indicate a peak temperature far from the red light band (red light wavelength is around 0.7 μm).

1.1.2 Various Instruments for Temperature Measurement

Today, many instruments are used as temperature sensors in industry. The following are typical means of measurement.

Glass thermometer. *Principle*: expansion and shrinking of a liquid (usually mercury or coloured alcohol) in a glass envelope due to temperature changes. It is widely used, being autonomous and not expensive. The possible measurement range with a glass thermometer is from -200 °C to $+600$ °C with an optimal accuracy of 0.03 °C. *Problems* (with respect to TNDE): point and contact measurement, high toxicity of mercury.

Thermocouple (invented by Seebeck in 1821). It represented 41% of the American market for temperature sensors in 1984 (Biermann 1988). *Principle*: an electric circuit formed of two different materials produces an electromotive force proportional to the temperature difference between the two junctions. Range: from near absolute zero up to 2750 °C. *Advantages*: one of the cheapest measurement methods; versatile, since the probe can be inserted in various ways while the reading unit is remotely located. *Drawbacks*: electric signals produced are small and prone to noise (consequently it is difficult to obtain accuracy greater than 0.5 °C); one of the two junctions must be at a known temperature; sensitive to drifts; relatively fragile; point and contact measurement may perturb the thermal field being measured; relatively long time to stabilize and to obtain a reading (in the order of 1 s) even if the mass probe is small.

Resistance (method invented by Siemens in 1871). It represented 25% of the American temperature sensor market in 1988 (Ortolano 1988). *Principle*: the resistivity of a conductor follows temperature variations. Advantages: reproducibility, excellent stability (for instance, platinum resistivity with a resolution of 0.006 °C is used as a temperature standard, Bailey 1988). Drawbacks: point and contact measurement, fragile, expensive (if platinum is used), difficult to obtain a good thermal contact.

Liquid crystals. *Principle*: cholesterol esters change orientation with temperature and reflect coloured light from red to violet when illuminated with white light. Depending on their composition, a 0.01 °C resolution can be obtained. *Advantages*: not expensive; sensitive to slight thermal variations; surface measurement possible. *Problems*: contact measurement; restricted sensitivity range (e.g. 5, 10 °C); necessity to prepare the surface before application as paint (Crabol 1987) or encapsulated in small balls or film (Cohen 1973); cleaning necessary after measurement to remove the crystals.

Radiation measurements. *Principle*: remote measurement of emitted energy. *Advantages*: since there is no contact, no thermal equilibrium is necessary between the object and sensor. *Problems*: emissivity and spurious reflections; more costly than other temperature measurement techniques.

This last category of instruments represents a practical example of the simple idea of using heat flux to inspect the internal structure of materials. Commercial availability of radiation measuring devices in the mid 1960s allowed the use of TNDE to spread rapidly. Although some applications of TNDE are possible employing contact methods, their uses are limited (Williams et al. 1983).

1.2 Active and Passive Approaches in TNDE

1.2.1 Passive Approach

Two approaches are generally recognized in TNDE, the *passive* approach and the *active* approach. The passive approach tests materials and structures which are naturally at a different (often higher) temperature than ambient. First investigations date from the 1930s: Barker in 1934 for fire forest detection, Nichols in 1935 to obtain temperature profiles during hot steel rolling. Important applications of the passive approach are listed below.

Production

Production is an important field of application for infrared thermography where abnormal temperature profiles indicate a potential problem which must be fixed in order to maintain the controlled process under valid operating conditions and before passing the product to the next production stage or to the customer. A few applications of special interest are briefly reviewed below.

Inspection of printed circuit boards (PCBs) to detect solder bridges and bad (overheating) components (Pau 1983; Williams and Fike 1987; Dresser 1990).

Quality control and seam tracking in arc welding (Wang and Chin 1986; Dufour and Maldague 1987; Nagarajan et al. 1988, 1989; Nagarajan and Chin 1990, 1992; Silk 1989; Gayer et al. 1990; Fuchs et al. 1991): it has been reported that recording of temperature patterns during welding allows tracking of seams. For instance, a nonuniform temperature pattern indicates misalignment of the torch with respect to the plate. Such infrared imaging is also capable of revealing some interesting parameters about seam quality, such as an estimation of the depth of penetration.

In the glass industry, recording of temperature profiles permits inspection of both sheet and hollow structures (such as bottles (Pajani 1987c)). For instance, temperature measurement and in particular measurement of the cooling rate in bottle manufacture enables the bottle wall thickness to be reduced, decreasing production cost (Wilson 1991).

In the production of metals, recording of temperature profiles enables monitoring of steel quality in continuous casting (Maldague et al. 1988) while hot spot detection in tank walls indicates production problems in nickel electrolysis (Pajani 1991).

In the paper industry, infrared thermography monitors quality in the production of high-gloss paper (Gaussorgues 1984a). In the cement industry, surveillance of rotating kilns is essential to maintain proper process conditions in cement production (Holmsten 1986).

Maintenance

Maintenance is an important field of application for infrared thermography where abnormal temperature profiles indicate a potential problem. A few applications of special interest are briefly reviewed below.

Inspection of turbine blades in jet engines (Ding 1985; Florin 1987) either in service or at the production stage. Active TNDE methods for turbine blade inspection at manufacturing stage are discussed in Chap. 5.

Thermal insulation of building envelopes (Ljungberg 1992), heated floors, furnace walls (Barre 1988). An example of such analysis is shown in Fig. 1.1 (see colour section): bad door insulation causes severe heat losses from this home. Investigation was performed during a cold winter night ($-27\,°C$) thus maximizing the indoor–outdoor temperature differential.

Estimation of liquid level in tanks (Smith 1987) or detection of buried pipe systems (Ljungberg 1992) is possible through infrared thermography. For instance, Waggoner (1991) reported detection of buried oil pipelines from data acquired by aircraft flown at an altitude of 600 m using Daedalus airborne scanners.

Inspection of electric installations is a big field of application for passive thermography (Safabakhsh 1989; Hurley 1992). For instance consider the situation of having a failure of one of the three transformers driving a three-phase motor: the motor will operate single-phase, causing potential motor failure. Early detection of transformer overheating by means of infrared thermography can prevent damage (Stovicek 1987).

Visualization of gas leaks: certain gases are invisible to the naked eye, thus causing potential safety hazards. This is the case for instance with hydrogen combustion which emits very little in the visible bands. This is a dangerous situation, especially in the aerospace field, where hydrogen is often used as a rocket propellant. Fortunately, observation is possible in the $8–14\ \mu m$ band because of the presence of water vapour, a product of combustion (Harper et al. 1990). Some gases stay invisible to this direct infrared visualization (e.g. air) and thus stay unnoticed, causing potentially dangerous situations. In some cases, observation is possible by illuminating the gas leak using a laser, or detection is possible if observation is performed against a heated (or cooled) background, revealing the shape of the gas plume. Another possibility for gas flow visualization is to place a filter in front of the infrared camera place, to detect one particular absorption band of the gas. Since the camera becomes sensitive to the radiation absorbed by the gas, the gas flow pattern is observed as a dark shape against a light (warm) background (Schmalz 1990). It is also possible to observe air turbulence by means of an electrically heated wire placed across the flow; because of the forced convection caused by the flow, observation of wire temperature is related to the flow behaviour, especially its velocity (Gartenberg and Roberts 1990). Rapid-scan spectrometers are also used for temperature measurement of gas and dust explosions, but this application is, however, outside of the scope of this book (see Cashdollar and Hertzberg (1982) for more details).

Medicine

Deployment of infrared thermography has been reported for evaluation of patients with disorders of the musculoskeletal system (Traycoff 1987, 1992) or for the evaluation of breast cancer where infrared thermography allows diagnosis at an earlier stage than with conventional mammography the presence of small tumours (Perl 1987). In fact, medicine is recognized as one of the first nonmilitary applications of thermography (Lawson 1956).

Monitoring of Road Traffic

In the case of monitoring of road traffic, infrared thermography enables moving road vehicles, which will be hotter than their surroundings (such as roads, buildings, (Jakowatz et al. 1987; Nandhakumar and Aggarwal 1988; Lu et al. 1992). For this application, infrared thermography can facilitate processing and interpretation of images.

Detection of Forest Fires

Detection of forest fires with infrared thermography using air-lifted and ground-based systems makes possible more efficient use of fire fighting units, while ensuring safety of forest areas through early detection of smouldering fires (Young 1985, 1992).

Agriculture and Biology

Infrared thermography is used in agriculture both for remote sensing (Lozano-Garciá et al. 1991) or for smaller scale applications such as for seed evaluation and recognition. Biological applications of the infrared have also been reported such as for behavioural observations of bats, *Eptesicus fuscus* (Kirkwood 1991).

Astronomy

In this case the approach cannot be, of course, *active*, and the techniques used are related to infrared spectroscopy (see Roellig et al. 1988; Bell et al. 1988; Krishnakumar et al. 1990).

Military

As mentioned previously, the military were among the first to envisage interesting opportunities for infrared vision systems in the two world wars, the main idea being to have a vision system capable of revealing the presence of potential targets in poor visibility conditions (e.g. at night or in fog). This is not surprising since heat released in the environment is a degraded form of energy with large entropy: it is a byproduct. In fact mechanical friction and combustion cannot escape heat dissipation. Heat is thus a loyal witness of human (and enemy) activities.

Nowadays, a lot of infrared research is still oriented in this direction with the addition of the synthesis of infrared images, which enables modelling and prediction of target signatures in particular conditions (Barnard and Boreman 1990), and also infrared satellite surveillance which allows, for instance, tactical missions in progress to be revealed from runways which remain hot long after aircraft have taken off (this technique was deployed in the 1991 Gulf war).

Another important military application of infrared is in the deployment of counter-measures (Waggoner 1991). For example, air-to-air detection of incoming enemy missiles or aircraft from their hot exhaust gases. In this case, infrared imaging and interpretation systems are air-lifted in the intercepting rocket. The AIM-9 Side Winder missile and Terminally Guided Submunition missile (TGSM) aimed at tank destruction are examples of such "intelligent" countermeasure device which sense "killer points" of the enemy target during the home-in phase based on prerecorded thermal signatures such as in the Target and Backgound Library System (TABILS). Notice that such systems are not always *passive* since in some instances, the target will be continuously illuminated with a laser beam while the interceptor vector locks itself on to the reflected radiation.

Summary

As we notice, the applications are numerous. In fact the worldwide market for infrared systems is estimated to be 500 units per year. The advantages of passive infrared inspection are many: reduced repair costs and lost of operations through early detection; reduced wear of equipment; reduced consumption of energy by having defective components fixed promptly; improved quality and operating conditions.

1.2.2 Active Approach

The active approach, unlike the passive approach, requires an external heat source to stimulate the materials to be inspected. This is, for instance, the case with vibro-thermography (McLaughlin et al. 1981) where, under the effect of mechanical vibra-tions (20 to 50 Hz) induced externally to the structure, heat is released by friction precisely at locations where there are defects such as cracks and delaminations.

Notice this distinction between passive and active approaches is not as clear cut as it seems. For instance, maintenance investigation can be performed by the active approach as well. For instance, moisture evaluation is reported by Grinzato and Mazzodli (1991). In this case they reported that presence of water activates undesired chemical reactions and thus the degradation process in monumental buildings. Due to the extremely high cost of restoration it is necessary first to assess the moisture problem. This can be done by uniformly heating the inspected wall from one side while observing the isotherm pattern on the other side: the thermal map recorded depends of the water content since heat capacity is greater for water than for dry material.

It is interesting to note that moisture evaluation of buildings, and roofs in parti-cular, has been reported by truly passive methods as well. In this case, thermographic inspection is performed at night and detection is based on the fact that moist areas retained day sunshine heat better than sound dry areas because of the high thermal capacity of water as shown in Table 1.1.

In this book, we put emphasis on *active* thermography, especially where a thermal pulse is applied to the material to be inspected. Following application of this thermal pulse, a measurement of the temporal evolution of the specimen surface temperature is performed with an infrared camera allowing subsurface defects to be revealed, as we explain below. Qualitatively, the phenomenon is as follows. The temperature of the material changes rapidly after the initial thermal perturbation because the thermal front propagates, by diffusion, under the surface and also because of radiation and convection losses. The presence of a defect reduces the diffusion rate so that when observing the surface temperature, defects appear as areas of different temperatures

Table 1.1. Thermal properties of some materials

Material	Specific heat $(\mathrm{J\,Kg^{-1}\,^{\circ}C)^{-1}}$	Density $(\mathrm{kg\,m^{-3}})$	Heat capicity $(\mathrm{J\,cm^{-3}\,^{\circ}C^{-1}})$	Thermal conductivity $(\mathrm{W\,m^{-1}\,^{\circ}C^{-1}})$	Thermal diffusivity[a] $\delta \times 10^{-6}$ $(\mathrm{m^2\,s^{-1}})$
Air (as defect)	700	1.2	0.8×10^{-3}	0.024	33
Aluminium	880	2700	2.4	230	95
Brass (65% Cu, 35% Zn)	380	8400	3.2	130	32
CFRP[b] ($\perp$ fibres)	1200	1600	1.9	0.8	0.42
CFRP ($\parallel$ fibres)	1200	1600	1.9	7	3.7
Concrete	800	2400	1.9	1	0.53
Copper	380	8900	3.4	380	110
Epoxy resin	1700	1300	2.2	0.2	0.09
Glass	670	2600	1.7	0.7	0.41
GRP[c] ($\perp$ fibres)	1200	1900	2.3	0.3	0.13
GRP ($\parallel$ fibres)	1200	1900	2.3	0.38	0.17
Lead	130	11300	1.5	35	23
Nickel	440	8900	3.9	91	23
Plexiglass™	—	1200	—	0.2	0.25
Porcelain	1100	2300	2.5	1.1	0.43
Steel (mild)	440	7900	3.5	46	13
Steel (stainless)	440	7900	3.5	25	7.1
Teflon™	—	—	—	0.42	1.59
Titanium	470	4500	2.1	16	7.6
Uranium	120	18700	2.2	27	12
Water	4180	1000	4.2	0.6	0.14
Zircaloy 2	280	6600	1.8	13	11

Adapted from Vavilov (1980, p 182); Reynolds and Wells (1984, p 43); Tretout, (1987, p 49); Touloukian and DeWitt (1970).
[a]Defined as $\delta = K/\rho C$, where K is thermal conductivity, ρ is mass density and C is specific heat.
[b]Carbon fibre reinforced plastic.
[c]Glass reinforced plastic.

with respect to surrounding sound areas once the thermal front has reached them. Consequently, deeper defects will be observed later and with a reduced contrast. In fact, the observation time t is a function (in a first approximation) of the square of the depth z (Cielo et al. 1987a) and the loss of contrast c is proportional to the cube of the depth (Allport and McHugh 1988).

$$t \approx \frac{z^2}{\delta} \quad \text{and} \quad c \approx \frac{1}{z^3} \qquad (1.1)$$

where δ is the thermal diffusivity of the material (cf. Table 1.1). These two relations show two limitations of the TNDE: observable defects will generally be shallow and the contrasts will be weak. An empirical rule says that *the radius of the smallest detectable defect should be at least one to two times larger than its depth under the surface* (Vavilov and Taylor 1982).

The detection phenomenon in TNDE is illustrated in Fig. 1.2. In this figure, the surface temperature evolution, following application of the initial thermal perturbation heat pulse, is plotted for a thin homogeneous plate (curve 1) and for the same plate when a delamination (air layer) is present under the surface (curve 2). On this logarithmic scale we notice the central portion of the temperature decay conforms

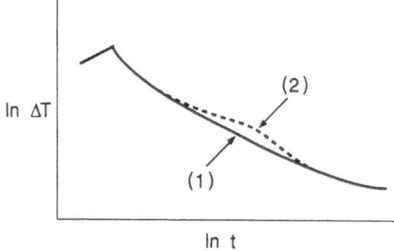

Fig. 1.2. Temperature evolution curve after absorption of a rectangular heat pulse. (1) Plate made of homogeneous material; (2) Same plate containing a subsurface flaw.

to a line of slope $(-1/2)$. In fact, for a semi-infinite medium, after absorption of a Dirac pulse, this temperature decay conforms to (Carlslaw and Jaeger 1959)

$$\Delta T = \frac{Q}{e\sqrt{\pi t}} \tag{1.2}$$

where ΔT is the temperature increase of the surface, Q is the quantity of energy absorbed and $e = \sqrt{K\rho C}$ is the thermal effusivity of the material with K being the thermal conductivity, ρ the mass density, C the specific heat and t the time. Table 1.1 lists thermal properties of common materials.

Utilization of a thermal pulse for the stimulation of the workpiece is very practical since all frequencies are tested simultaneously (flat frequency spectrum of a Dirac pulse) but there is reduced sensitivity. Another possible approach makes use of periodic thermal cycling of the specimen. In this case individual frequencies are tested separately. This last approach is, however, time consuming and not always practical for automated inspection in real time, though it permits great sensitivity of measurement.

Another procedure, for the detection of surface cracks, is to have the thermal front propagating along the surface (Burger and Babak 1985; Vavilov 1990). The stimulation can be achieved with the specimen being suddenly brought into contact with a thermal mass (Fig. 1.3). In this case surface crack of higher thermal resistance oppose to the passage of the thermal front leading to surface temperature differentials detectable by the infrared camera. Surface propagation of a thermal front can also be achieved by vertical immersion of one section of the specimen in a hot (or cold) water tank while observation is conducted on the non-immersed section with the thermal front propagating bottom-up (Milne and Carter 1988). Although this approach does

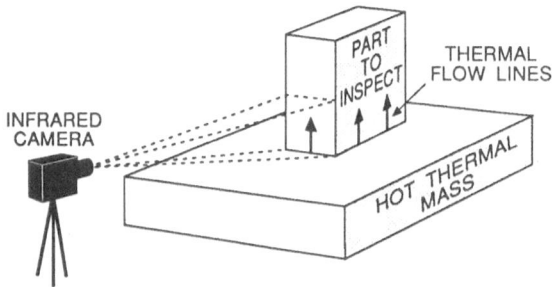

Fig. 1.3. Heat front propagation along the surface after sudden contact of the sample with a thermal mass of different temperature.

not lead directly to repetitive inspection on the plant floor, it can be of great interest for specific applications.

In the case of the pulse heating approach, different configurations are possible as illustrated in Fig. 1.4 (Fig. 1.5 shows thermogram aspects). In this figure, we present three methods: (a) *Point heating*, e.g. by means of a laser beam or a focused arc lamp: a rather uniform heating is obtained from point to point; the drawback is the necessity to scan the beam if a surface has to be inspected, it is thus a slow method; (b) *Line heating* (using a line infrared lamp, a heated wire, a line of air jets): good uniformity and fast inspection rate are obtained because of the lateral scanning; (c) *Surface heating* (using cinematographic spots, incandescent bulbs, flash lamps, fast laser scanning): it is difficult to obtain uniform heating, and the configuration can be either static or mobile.

Position b of the detector (Fig. 1.4b) with respect to the heating source is of great importance since, as we have seen (Eq. (1.1)), the optimal time of observation $t_0 = b/v$ is a function of the depth of the defect under observation. Thus point and line heating

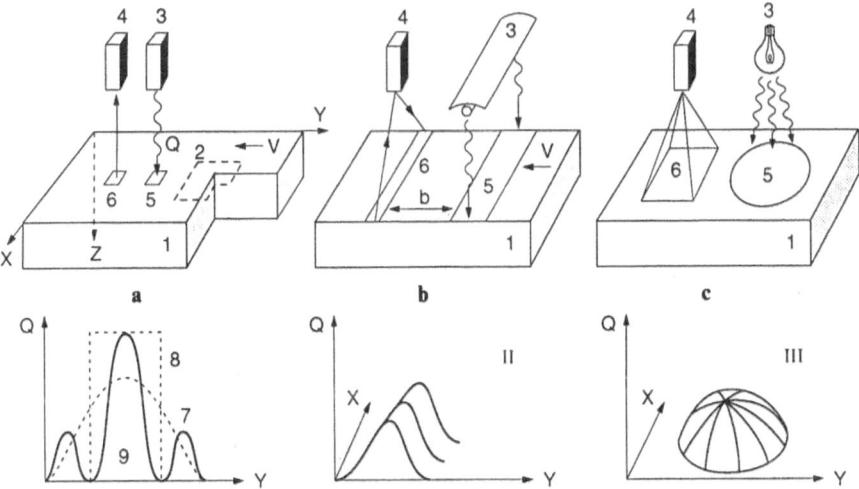

Fig. 1.4. Different configurations for TNDE inspection: **a** by point; **b** by line; **c** by surface. I, II, III: energy distribution by the thermal source. 1: sample; 2: defect; 3: thermal source; 4: infrared detection system; 5: heated area; 6: observation area; 7, 8, 9: heating distributions–Gaussian, uniform and circular, respectively.

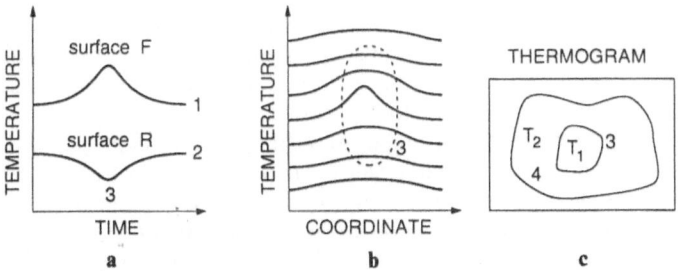

Fig. 1.5. Shape of the thermograms for **a** point heating; **b** line heating; **c** surface heating. 1, 2: temperature profiles (case of a plate, heating of F: front surface, R: rear surface); 3: zone with defect; 4: heated area.

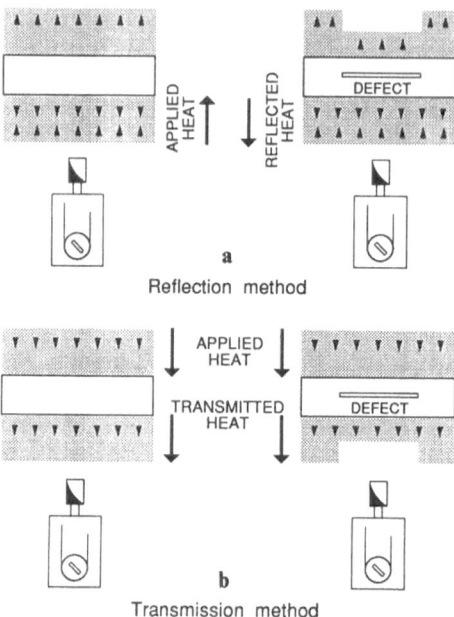

Fig. 1.6. Observation methods: **a** in reflection; **b** in transmission. Note that the thermal front propagation can be either cold or hot.

methods are favourable only when defect detection is limited to cases where defect depth is known and constant. For instance in the case of bonded assemblies where defects are located between layers of known thickness, presumably at the bonding interface.

Two methods of observation are possible (Fig. 1.6): (a) *reflection*, where both the thermal source and the detector unit are located on the same side of the workpiece; (b) *transmission*, where the thermal source and the detector unit are located on opposite sides of the workpiece. These two observation methods do not offer the same possibility of defects detection. With reflection, a greater resolution is obtained, but the thickness of the inspected material is small (Reynolds 1985). With transmission, a greater thickness of material can be inspected but the depth information is lost and since the resolution is weak, it is necessary to use more sensitive detection equipment (Sayers 1984). Moreover, observation is not always possible with the transmission method especially for complex structures made of multiple layers (e.g. honeycomb panels). Generally, the reflection approach is good for the detection of defects located close to the heated surface while the transmission approach permits one to reveal defects located close to the rear surface.

It is important to notice that a cold front (heat removal) will propagate in the same fashion as a hot front (heat deposit) inside the material. In some inspection situations, this *cold* approach is more economical. For instance for the inspection of a component already at an elevated temperature (with respect to ambient temperature) due to the manufacturing process such as for the curing of bonded parts. Moreover, it may be attractive to preheat (if necessary) the surface with a low rate of energy deposition (by means of electric heating wires for example). This approach avoids relying on high intensity local heat sources which are more costly, age rapidly and are potentially dangerous to the surface because of the high thermal stress they induce. Cold thermal sources are also advantageous since they do not generate thermal reflective noise which can perturb the measurement (see Chap. 4 for more details on this cold approach).

The diversity of thermal stimulation sources is very great; authors have reported the use of boiling water, snow, incandescent lamps, lasers, plasma arcs, inductive heating, warm or cool air projection, radiative heating elements such as wire coils, flash lamps, and aerosol sprays such as spray of cold liquid nitrogen vapour.

Progress can be expected in TNDE with the availability of uniform high energy density, compact thermal stimulation sources which, moreover, would minimize parasitic radiation. It is also important to realize that the stimulation source must stay nondestructive, not damaging the inspected surface either chemically or physically.

1.3 New Materials

1.3.1 General Considerations

On Materials

In our modern world, numerous new materials are now employed while the traditional supremacy of iron (5% of the Earth's crust) and aluminium (8% of the Earth's crust) is reduced. Among these new materials, there are multilayer laminates, carbon fibre reinforced plastics (CFRP) also called graphite epoxy components, honeycomb panels, ceramic coatings, thermal barriers (Mansour 1983), metallic matrix composites, new alloys such as Al–Cu–Li and Al–Li–Cu–Mg (Danjoux et al. 1987; Woodward 1991). These new materials come often from the aerospace and military industries. For instance new ceramic coatings enable weight reduction, increased efficiency and substantial fuel consumption savings of jet engines operating at higher temperatures (Feest 1988). These new materials, even though they cost around, $2.50/lb against $0.40/lb typically for traditional materials (approx. prices in 1992 US dollars), are finding many new applications in the motor vehicle, leisure boat, bicycle, shoe and building industries. Such new materials are often less, or not prone to corrosion, they require fewer finishing operations, and their use also enables a reduction in the amount of parts needed to complete an assembly. For instance, fabrication of the boot lid of the Fiat Tipo necessitated 20 parts using steel while the plastic version is made of only two, with a substantial reduction in assembly time.

Probability of Defect Detection Reliability and Inspection Programmes

The physics of damage and mechanical theory of failures brings us to the concept of tolerance to damage. The idea is to be able to predict part behaviour as a function of local constraints; it thus becomes possible to define acceptable defects (Loubet 1987; Agam et al. 1988). In this respect, Fig. 1.7 gives the probability of defect detection as a function of defect size. This is an experimental curve valid for a specific method of inspection. On this curve, we can see that the ratio between the threshold of detection and a reliable detection value can be as much as ten. It is thus important to have an

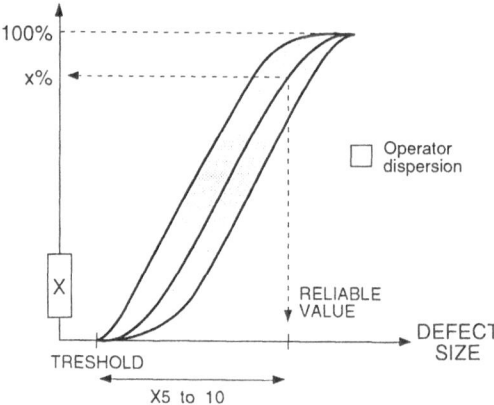

Fig. 1.7. Probability detection curve. The region marked with a cross indicates where false readings occur.

inspection strategy which allows us to distinguish between superficial tolerable defects and unacceptable ones.

Important considerations concern the reliability of a given part or systems (Madrid 1990). In fact, it is well known that reliability can be expressed as a *bath curve* (Kraft and Wing 1981). In the early life period of a component or a system, the failure rate is high. Next, as early failures (primary failures) are replaced, the components or system settle down to a long relatively steady period at lower and constant failure rate. This period is often referred to as the *useful life period*. Later, in the *wear-out period*, components or system rapidly deteriorate and the failure rate rises again. The wear-out period can be avoided if components are replaced before they reach this period (inspection is important to determine when to replace components). As an example, the human life cycle follows this pattern with the high infant mortality rate which lasts up to about the tenth year followed by a period where the majority who survive live to an old age. In the seventieth year or so, the curve rises as people begin to die of natural causes. The death rate increases further until all the individuals in the group are dead. Of course, during the *useful life period*, the failure rate can increase if the system or component is disturbed by accident, unusual operations, external conditions or bad maintenance. This discussion implies that if a (thermographic) inspection programme is set up as part of a maintenance programme, after some time, it should not continue to *produce results*, that is to report findings (detection of defects). This situation does not mean, however, that the inspection programme then needs to be stopped; on the contrary, it means that the maintenance is sufficient and inspection is still needed to double-check (Lucier 1991).

1.3.2 NDE Techniques for New Materials

Other Methods of Detection

The TNDE method is particularly well suited for the inspection of bonded assemblies, composites and honeycomb structures which are difficult to inspect with other traditional NDE methods, the so-called *big five* (Smith 1987) which are ultrasonics,

X-ray radiography, eddy currents, dye penetrant and magnetic flux leakage (see for instance Cielo (1988, 1989) for a more complete discussion of NDE methods).

Ultrasonics is probably the most used NDE method. Ultrasonics offers interesting possibilities but is characterized by (slow) point by point inspection rates (Hudson 1985). The principle of ultrasonic NDE is based on a pulse–echo scheme where the defect reflects (or absorbs – in transmission) ultrasonic waves of high frequency (1 to 25 MHz). The necessity of contact with the inspection zone in order to ensure good signal propagation between the transducer and the workpiece is one of the strong disadvantages of this method. This contact can also be achieved using water jets or by immersion in a water tank. These last two methods are more easily automated than the manual procedure using wedge transducers and gel coupling, but exposure of the workpiece to water is undesirable in some cases. New noncontact techniques of generation and detection of ultrasound by laser are very promising (Jen et al. 1985; Cielo et al. 1986b; Rousset et al. 1986; Maldague et al. 1986b; Monchalin 1986). The principle is as follows. The thermal stress induced by the generation laser initiates the ultrasound propagation. Reflected or transmitted mechanical ultrasonic waves locally deform the workpiece surface and these mechanical deformations (a few nanometres in amplitude) are picked up by the detection laser through interferometry methods.

X-ray radiography (Mallick 1986) or gamma radiography (for deeper penetration) allows one to detect some kinds of defects (voids, air layers, presence of foreign materials) because of the differential absorption of the radiation due to density differences. This NDE method has some problems related to health security of nearby personnel, and it is not efficient in detecting thin air layers (since attenuation of X-ray beam is insufficient in this case). Moreover, real-time radiography achieves low resolution compared with film radiography (Mengers 1986).

Other NDE methods are used on new materials with variable success; holographic interferometry (Monti 1986), positron annihilation (Smith 1987), acoustic emission (Vacelet 1984) and the *mirage* technique, for instance. The mirage technique allows extremely sensitive measurements: the sample surface is heated periodically by a modulated light beam and the deviation of a probe beam propagating in a parallel direction, just above the heated areas, is related to periodic temperature of the sample. In turn, the sample temperature will reveal specific sample characteristics. For instance, it is possible to determine optical losses in mirror coatings with a resolution of a few tens per million or to detect very thin air layers (with a resolution of $\sim 0.1~\mu m$) in metallic components (Charbonnier et al. 1989). Drawbacks of the mirage technique include sensitivity to mechanical vibration, to beam misalignment and to beam optical noise, which necessitates careful design of the inspection set-up.

TNDE and New Materials

In the aerospace industry where inspection is of prime importance at all the fabrication stages, mainly for security reasons, there is a need for a fast and economical inspection technique. For instance, the Lockheed C-5 transport aircraft is made of some $3400~\text{m}^2$ of bonded structures, completely inspected using infrared thermography (Cohen 1973). Another example of the generalized use of composite materials is the new Boeing 777 (first aircraft scheduled for delivery in May 1995) in which 10% of the total weight of the structure will be made of graphite epoxy composites, compared with a mere 3% in the case of other Boeing aircraft. In fact, infrared thermography has always been quite active in the aerospace field as the numerous papers on the subject demonstrate, for instance the work of Balageas and Luc (1986); Balageas et al. (1987c, 1988, 1990, 1991); Quinn (1988); Tretout (1987); Tretout and Marin (1985); Tretout et al. (1991). One reason for the good acceptance of TNDE in aerospace industry is that TNDE is well adapted to the inspection of low thermal conductivity materials such as CFRP or ceramics due to the low scanning rate of thermal imagers (Tretout et al. 1991). However, for the inspection of high thermal conductivity workpieces, the low scanning rate limitation can be overcome through image processing of the whole infrared image sequence recorded after application of the thermal heating pulse. The procedure can be summarized as follows. The experimental decay is computed on a pixel-by-pixel basis, taking into account the low scanning rate limitation and comparing with theoretical decay modelled after a similar structure as the one inspected (see Chap. 2 for more details on modelling). Delpech and Balageas (1991) reported such investigations on metal–metal bonded assemblies, two-layer bonded composites and delamination in CFRP panels.

Active TNDE is also suitable for inspection of advanced new materials in the nuclear industry (Green 1968; Lewak 1992). For instance, Gitzhofer et al. (1987) reported the study of thermal barriers projected by plasma.

Although in this book we concentrate on thermal bi-dimensional imaging, it is important to notice that point heating (Fig. 1.4a) offers very attractive characteristics for specific applications, for instance on advanced ceramics (Cielo et al. 1986c). Since heating can be performed with a laser beam, short heating pulses approximating the delta (impulse) function can be achieved (e.g. in the 15 ns range with a $1.06\,\mu m$ Nd:YAG laser in the Q-switch mode) while the detection with a focused single detector is not limited to the 25 or 30 Hz video rate. Consequently, fast thermal events can be caught, while wavelength scanning with a broadband single detector and appropriate filter is also possible, for instance to select a particular thermal emission wavelength. This point thermal wave method can be applied, for instance, to the measurement of clearcoat thickness in metallic paint systems of car bodies (Imhof et al. 1991). Surface is pulsed-laser heated (at 532 nm) and the peak delay time of the emitted thermal radiation picked by the single detector is related to the clearcoat thickness $(20\text{–}50\,\mu m)$. In this application, it is important to select an excitation wavelength that is transmitted by the coating and absorbed by the substrate so that from substrate–film interface to the top surface, the reflected thermal wave diffuses through the film thickness: the thicker the coating, the longer the travelling time, thus delaying the observed temperature peak.

Thermographic imaging of surface finish defects in coating on metal substrates is also possible by heating the coating system from the back slightly above ambient temperature (up to 20–30 °C) with a heating pad and viewing it with the infrared camera. Surface finish defects (craters or protrusions) are visible due to coating thickness variation between defect and sound areas, since coating thickness affects the thermal radiation emission/reflection/transmission properties of the coating system; resolutions of about $20\,\mu m$ are reported (Bentz and Martin 1992).

1.3.3 Bonded Assemblies

We will now study more specifically a few new materials where TNDE has been applied with success. In the case of other materials such as metals or ceramics, it is worthwhile to say that, as a general rule, TNDE is sensitive to voids, inclusions, cracks, delaminations and to the presence of water.

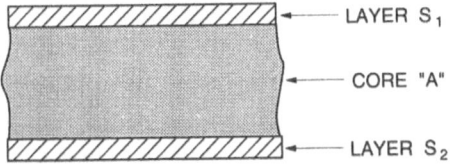

Fig. 1.8. Cut view of a sandwich structural panel.

Sandwich structural panels (Sanmartin 1988) consist of two plates S_1 and S_2 (Fig. 1.8) of high mechanical strength (e.g. sheets of Al, Ti, steel, etc.) and of a core A of low density (foam, honeycomb, balsa, wood etc.) used as a spacer element between the plates. These materials are employed to build structures having great strength, low mass and good thermal insulation capability as well as good fire protection.

Bonded assemblies offer interesting advantages (Vacelet 1984): better distribution of the mechanical loading, superior resistance to fatigue with respect to the more traditional point-by-point welding and riveting anchorages, lengthened life span, air- and watertightness, improved resistance to the corrosion, ability to ensure electrical or thermal insulation.

The quality (strength and durability) of a bond depends mainly on the interaction of the adhesive with the adherent (bonding surfaces). The quality preparation is thus essential (McNamara and Ahearn 1987; Dickstein et al. 1991). Bonding is particularly affected by ambient temperature and humidity at time of bonding. If all the fabrication parameters are not perfectly controlled, different types of defect are likely to occur such as (1) lack of adhesive (bubbles, air layers, foreign materials), (2) cohesion defects (breaking within the adhesive), (3) bonding defects (breaking at surface/bond interface). Defects of types (2) and (3) are very difficult to detect whatever NDE method is used (Segal et al. 1979; Cielo et al. 1985a). TNDE offers good possibilities for detecting defects of type (1).

1.3.4 Graphite Epoxy Structures

Graphite epoxy structures consist of a matrix of carbon fibres embedded in epoxy resin (up to 32% of resin (Sadat 1988)). They are widely used thanks to their excellent mechanical properties. These properties enable moulding in a wide variety of shapes, thus reducing the quantity of parts to be assembled and consequently lowering the fabrication costs (Dorey 1988). Moreover, since the anisotropy is easily controllable (it depends of the orientation of the individual plies), the designer is free to adjust the strength and the mechanical resistance with respect to the envisaged mechanical loading of the part (Lang 1988). Typical defects are voids, inclusions, areas with unbalanced resin content, badly cured areas, surface cracks and broken fibres. Cutting and drilling tend to induce delaminations which occur at the edges and which can alter fibre orientation, thus causing a loss of rigidity. Also of importance is damage caused by impact, inducing delaminations and cracks propagating in a parallel direction to the plies. These delaminations take place especially within the more disorientated plies. Often, as we will see in Sect. 4.2.1, the damage is barely visible on the impact side, but considerable on the other side. This is often called *blind side impact damage* (McLaughlin et al. 1981). In the case of graphite epoxy structures, TNDE offers good possibilities for detection of delamination, inclusions and impact damage, as well as for the evaluation of fibre content and orientation.

1.4 Detectors for Infrared Imaging

More detailed studies concerning detectors can be found elsewhere (see for instance Hudson 1969; Gaussorgues 1984b; Coester 1988; Norton 1991; Maldague 1992). In this section we will review briefly what knowledge is essential with respect to understanding the following chapters.

One of the first devices used to produce infrared images was the *evaporograph* (Czerny, 1929) whose sensitivity in the infrared is due to the differential evaporation of a thin film of oil on a dark membrane. Differences of reflectivity with visible light on the dark membrane allow the visualization of the infrared image. Exposure time varies from a fraction of a second to a few minutes depending on the temperatures of the objects observed. In 1840, Sir John Herschel (son of Sir William Herschel, Sect. 1.1) was the first to produce such an infrared image.

Since then, things have changed. Detectors can be divided into two large families: *thermal* and *photonic* detectors. Figure 1.9 shows the spectral response for the most common infrared detectors. The spectral atmospheric transmission is also plotted. In this figure, the two major spectral windows of the atmosphere used in infrared thermography ($3-5\,\mu$m and $8-14\,\mu$m) are indicated. For these bands, radiation measurement is possible without too much attenuation because they match atmospheric transmission well.

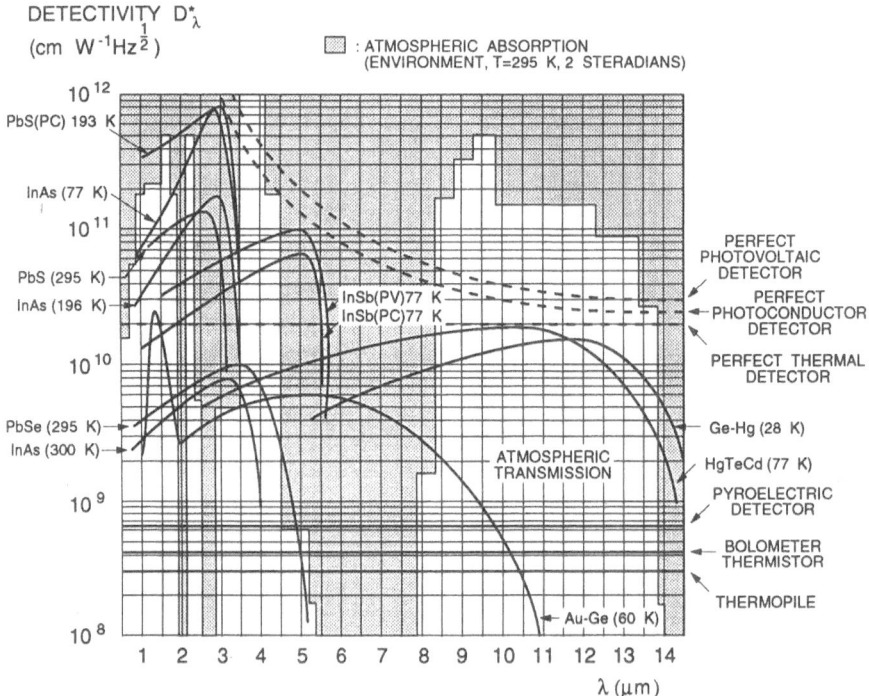

Fig. 1.9. Spectral detectivity curves of infrared detectors. Atmospheric absorption bands are also indicated.

1.4.1 Thermal Detectors and Cameras

In thermal detectors, infrared radiation is absorbed and produces a temperature change in the detector itself. Any physical property sensitive to temperature can be used. We have already reviewed the glass thermometer, thermal resistances, thermo-couples and liquid crystals which are based on this principle (Sect. 1.1.2). In infrared thermography, a widely used thermal detector is the pyroelectric vidicon tube in which a pyroelectric target releases electric charges because of localized heating (Goss 1987). Scanning of this target by an electron beam allows the collected image to be read. Pyroelectric targets are made of certain ferroelectric crystal materials such as triglycine sulphate (TGS) having a high pyroelectric coefficient (p). This kind of detector is widely used (Stillwell, 1981; Vavilov 1984; Burgess et al. 1985; Monti 1986; Dixon 1988; Calais and House 1990).

Advantages are

1. Wide spectral response (limited only by the entry optics and the presence of an interference filter.
2. No cooling required.
3. Reasonable cost (price of a camera $\sim$ \$15 000).
4. Compatibility with standard TV signals, as the method used to scan images in TV tubes and in pyroelectric tubes is the same. In fact, changing the electron beam scanning rate suffices in obtaining another video format.
5. Since there is no mechanical scanning involved in the image formation process, a greater reliability is gained.

Disadvantages are

1. Necessity to use a mechanical chopper synchronized with the electronic reading beam since the pyroelectric target is only sensitive to changes of radiation. This aspect can, however, present two advantages: the chopper produces an alternating signal which can be amplified more easily than a DC signal; moreover, the chopper can be used as an absolute temperature reference to calibrate the instrument automatically (Liddicoat and Marsi 1988). Instead of using a rotating chopper, the camera can be passed over the scene. Without chopper or panning mechanisms the instrument can be used to observe moving vehicles or intruders automatically, being sensitive only to temperature changes.
2. The thermal image is of average quality because of the image flickering induced by the chopper (this effect can, however, be corrected with appropriate electronics).
3. Small range of observable temperature.
4. Restricted life span ($\sim$ 5000 h) and rapid ageing of the tube. If a given pattern is observed repeatedly, the tube starts to "memorize" it, causing a ghost image to remain permanently observable.
5. Detectivity only about half as good as photonic detectors.
6. Low stability, strong nonlinearities and spatial nonuniformities of the image imply periodic recalibrations.
7. Since the detector heats itself, the image tends to smear laterally by conduction on the detector surface.

Manufacturers include:

Electrophysics, Nutley, NJ, USA

FJW Industries, Singapore

Hamamatsu, Japan

ISI Group, Albuquerque, NM, USA

Image Technology Methods, St Waltham, MA, USA

Insight Vision, Worcestershire, UK

Mikron, Wyckoff, NJ, USA

Xedar, Boulder, CO, USA

1.4.2 Photonic Detectors and Cameras

In photonic detectors, the energy is absorbed and affects atomic states and free electrons within the semiconductor. In this way, the energy of an incident photon can be sufficient to release an electron, augmenting the density of the free electrons and thus increasing the electrical conductivity of the detector (in the case of photonic photoconductor detectors) or the output voltage (in the case of photonic photoelectric or photovoltaic detectors). Since a given amount of energy is necessary to release free electrons and charges, the detector response will depend of the photon wavelengths. Since photonic detectors are not based upon a temperature change, their response time will be faster than for thermal detectors. Common materials used in phototonic detectors are: Si, InAs, InSb, HgCdTe (also called cadmium mercury telluride (CMT)).

There exist three major types of infrared cameras built around photonic detectors: (1) mono detectors with associated mechanical scanning for image formation, (2) cameras using a SPRITE detector associated with mechanical scanning for image formation (the SPRITE – Signal PRocessing In The Element – was originally developed in the UK at the Royal Signals and Radar Establishment by Dr. Ted Elliot (Elliot 1981; Bell 1991; Whitlock et al. 1991)), (3) cameras using a focal plane array (FPA) type of detector with associated electronic scanning of the image. FPAs are reviewed in more detail at the end of this section.

Advantages are

1. Excellent detectivity, close to the theoretical limit.
2. Temperature calibration relatively easy (single detector) and reproducible in type (1).
3. Type (3) is robust since there are no moving parts.
4. Types (1) and (2) are excellent for quantitative measurements.

Disadvantages are

1. High cost, typically $\sim \$60\,000$ per unit.
2. Necessity to cool the detector down to a cryogenic temperature. See paragraph below about cooling methods.
3. Slow response time: infrared cameras operating at scanning frequency of 50 or 60 Hz (TV standard) cannot catch fast thermal events, for instance in high thermal conductivity materials such as aluminium. This can prevent TNDE being deployed for some applications. As mentioned above (Sect. 1.3.2), special image processing techniques can relax this constraint.

4. Temperature calibration can be a problem in types (2) and (3), especially if the number of detector elements is large and if their individual response is not very uniform. The calibration process may slow the image processing if correction has to be applied on an individual detector basis. See paragraph below about temperature calibration of infrared cameras.

5. The electro-mechanical scanning mechanism in types (1) and (2) may not prove robust enough in harsh industrial environments, moreover this mechanical scanning may also introduce distortions in the video signals such as the presence of unwanted patterns (Yoder 1968).

Some manufacturers of types (1) and (2) are

AGEMA, Daneryd, Sweden
Inframetrics, Bedford, MA, USA
FLIR Systems, Portland, OR, USA
Rank Taylor Robson, UK

Some manufacturers of type (3) are

Amber Engineering, Goleta, CA, USA
David Sarnoff Research Center, Princeton, NJ, USA
EG&G Reticon, Sunnyvale, CA, USA
Fairchild Semiconductor, Milpitas, CA, USA
Mitsubishi, Japan

Focal Plane Arrays

Focal plane arrays (FPA) are of recent development (Aguilera, 1987; Jost et al. 1987; Shepherd and Moorey 1987; Bahraman et al. 1987; Pellegrini 1987; Kimata et al. 1988; Ravich 1988). In 1984, models with 244×160 detector cells were available; in 1986, 256×256; and since 1987, 512×512 is standardly available from many manufacturers. There exist many different fabricating technologies such as:

Schottky barriers. Based on the well established silicon technology, platinum silicide (PtSi) technology with response in the range 3–$5 \, \mu m$ is the most established.

Superlattices. Alternating layers of various semiconductors of different thickness allows tuning to the wavelength of radiation that will be absorbed: photoconduction occurs in a narrow range of wavelengths. One of the most promising technologies is GaAs/GaAlAs first proposed by Levine et al. (1987). Typical cutoff frequencies are between 6 and 11 μm.

Intrinsic photon detectors. These detectors are in fact arrays of the photoconductive or photoelectric (photovoltaic) detectors we have discussed above. Photoelectric (photovoltaic) detectors are more useful since they generate charges spontaneously under illumination by incident radiation. Hybrid technology, where detector layers are fused or glued on silicon readout circuits is attractive because the silicon process is a well-established fabricating process. However, mismatch between the thermal coefficient of expansion sometimes causes a problem at cryogenic temperatures. Such hybrid arrays are fabricated with HgCdTe detectors in sizes of up to 128×128, 8–$12 \, \mu m$ (Baker et al. 1990) and 256×256, 3–$5 \, \mu m$ (Gubala et al. 1989). An alternative fabricating technique is to grow HgCdTe cells on GaAs buffers, themselves grown on silicon substrate containing bipolar preamplifier transistors and readout circuits. Advantages of this monolithic configuration include improved uniformity, reduced $1/f$ noise, and higher-operating temperature (Ballingall 1990; Nelson et al. 1991). Monolithic FPAs where HgCdTe diodes are directly grown on Si could also yield to large arrays (Zanio 1990). HgCdTe, InSb, PbSnTe and InGaAs are the more common types of intrinsic detectors, with newer compositions such as HgMnTe and HgZnTe developing.

Cooling Methods for Infrared Cameras

The easiest method is to make use of *liquid nitrogen* (LN$_2$, $-196\,°C$) poured in a small vessel (called a dewar) held in the camera itself while the detector is bonded directly to the cold wall. This method is cheap but not convenient for some applications since the dewar must be refilled every 3 h or so depending on its capacity, generally a few hundred millilitres.

Another method employs thermoelectric *Peltier elements*. These semiconductors are made of two dissimilar metals and have the property to transfer heat from one junction to the other depending on the direction of the current flow. Up to seven stages of Peltier elements can be stacked upon each other in order to reach a temperature differential of up to about $-125\,°C$. Peltier effect cooling is thus less efficient that LN$_2$ cooling and infrared images obtained from Peltier-effect-cooled imagers tend to be slightly noiser.

A popular cooling method is *Joule–Thompson gas expansion* for which the quick expansion of a high pressure gas (such as nitrogen or argon) produces, after a few minutes of operation, droplets of liquid nitrogen or liquid argon ($-187\,°C$) at the tip of the expansion nozzle (on which the detector is fastened). This mechanism permits greater autonomy than dewar operation; it is, however, noisy, and the gas tank may be cumbersome.

Finally, cryostats using a *closed Stirling cycle engine* can also be employed for cooling. This machine cools through repetitive compression and expansion cycles of a gas by means of a piston: it compresses gas at a low temperature and allows it to expand at a high temperature, as in household refrigerators. Typical input powers involved are around 4 W for 150 mW of cooling for a small engine (0.5 kg) while a large engine (2 kg) will deliver 1 W of cooling for 40 W of input power. This last method is advantageous for operation along the product line since it does not require any refilling.

Temperature Calibration of Infrared Cameras

Temperature calibration is done as follows. The infrared array (e.g. FPA) is exposed to a scene of uniform temperature and an image is recorded in computer memory. This process is repeated for different scene temperatures. Next a calibration function, either linear or of higher order for nonlinear response to the photonflux, is computed pixel by pixel (Maldague et al. 1991b). For linear functions, two coefficients are obtained for every pixel; this is the so-called "two-point method": gain and offset. Second order polynomials require three parameters and more coefficients are needed for multiple-point correction. In single-point compensation the average of several images recorded over a uniform background with a defocused lens is subtracted (in real time) from live video to remove nonuniformity. This process is, however, less efficient than other techniques such as two- and multiple-point methods (Tower 1991). During normal operation of the array these computed coefficients serve to correct images. More details on temperature calibration are given in Sect. 3.3.

1.5 TNDE: Pros and Cons

In NDE, no single technique offers a solution to all problems (McLaughlin et al. 1981). Often it is necessary to combine different techniques at different inspection stages. For instance, infrared thermography can first be applied to sort the parts rapidly on a pass–fail criterion before inspecting rejected parts more closely with ultrasonics in order to pinpoint the exact extent of the detected problems. Every NDE technique has its own strengths and weaknesses. We will compare these for TNDE.

Strengths and Advantages

1. Fast surface inspection, limited mainly by the power and the geometry of the thermal stimulation device. It has been reported that in some instances, thermography allows inspection as good as with ultrasonics or X-rays in only 20% of the time required by these techniques (Chambers 1984).

2. No physical contact is required, either for stimulation (by the thermal source) or for detection (by the infrared camera).

3. Ease with which the TNDE apparatus can be employed with minimum operator intervention.

4. Possible utilization even if access is restricted to one side only of the inspected component. If desired, inspection is also possible on both sides in transmission, as seen in Fig. 1.6.

5. Security: no harmful radiation is emitted, while the restricted surface heating of 5 to 10 °C generally suffices for the TNDE inspection and does not damage inspected components.

6. Numerical computation, required to solve diffusion equations of heat transfer through finite difference modelling, is accomplished with relative ease (see Chap. 2).

7. Ease of interpretation of thermograms, and ease of comparison with analytical models which appear as images.

8. Thermograms can be archived easily (for instance on magnetic media for long periods of time, e.g. during the life span of a component). Such archived data can even be used to prolong the active life of components (if records reveal sound structures without major crack growth, for instance).

9. Great versatility of applications: TNDE can be applied to almost all materials and their composites. Defect detection from a fraction of a millimetre in size and at a depth of up to 13 mm have been reported (Quinn et al. 1988).

10. TNDE is sometimes a unique tool to obtain information about internal structure of a component.

Problems and Weaknesses

1. Difficult to obtain a uniform brief high energy deposit on a large surface (Winfree and Welch 1989).
2. The variable emissivity of a surface may provoke perturbing contrasts at the interpretation stage of thermograms. As we will see in Chap. 2, low emissivity can also induce problems of parasitic reflections from nearby hotter bodies.
3. Cooling losses by convection and radiation also generate perturbing contrasts.
4. Over long distances (greater than a few metres), absorption of the infrared signal by the atmosphere must be taken into consideration. This is not necessarily a straightforward task. Various atmosphere models can be considered: Schott and Biegel 1987; Jarem et al. 1984; Shushan et al. 1991.
5. The transitory nature of the thermal contrasts on surfaces of high thermal conductivity materials imposes an inspection rate sometimes difficult to sustain with conventional video rate infrared equipment. For nonmetal materials of lower thermal conductivity, the inspection time window is more flexible: 10 to 100 s typically.
6. Relatively high cost of infrared equipment, especially if quantitative measurements are required (up to $150 000 for state of the art infrared cameras, whereas manual thermal inspection can cost nothing).
7. Requirement for a direct straight viewing corridor between the inspected surface and the detector. However, folded paths with high reflective mirrors (such as first surface gold mirror, Fig. 3.8) or the use of infrared fibre optics bundles relax this requirement (Kaplan 1987).
8. High rate of false readings due to the small signal to noise ratio (SNR): the inspection proceeds with detection of temperature differentials of a fraction of a degree over a background temperature of about 300 K. Signal processing techniques are thus required to limit sources of noise, extract the useful information and present them in the most appropriate fashion. Of course, referring to both Fig. 1.7 and Eq. (1.1), we come to the conclusion that detection will be impossible for some very small defects located very deeply and also in the case of close cracks which do not offer any thermal resistance to the lateral propagation of the thermal front (Fig. 1.3).

Conclusion. In this chapter, we have presented TNDE in a general context. In the following chapters, we will study in more depth aspects only briefly discussed here. We will start in the next chapter, with some theoretical aspects of radiometry and heat transfer modelling which are fundamental to quantitative thermographic analysis.

Chapter 2
Theoretical Aspects

2.1 Radiometry

Radiometry is concerned with the measurement of radiated electromagnetic energy (see Spiro (1990) and Beaudoin and Bissieux (1992) for a more complete coverage of this matter). Contrary to measurement of time, frequency or voltage, where resolution of the order of 1 part in 10^{10} or even 10^{11} can be obtained, in radiometry a resolution of only a few percent is generally the best that can be envisaged and only in very careful experimental conditions. Reaching the fraction of a percent or even the percent itself is not an easy task. The mirage measurement (Sect. 2.3.2) is, however, an exception to this situation, with reported temperature measurement sensitivity of up to 10^{-8} K in rigorous laboratory conditions (temperature measurement of liquid CCl_4 (Lepoutre and Roger 1987).

The poor performance of radiometric measurement is due to numerous factors (Nicodemus 1967): large time intervals with respect to one period of the radiation frequency; large distances involved with respect to the wavelength; dissipation of the power radiated in all space; great variation in the power radiated depending on the wavelength, position, direction and polarization.

The liberated photonic energy W, following an increase of temperature (increase of molecular agitation) is expressed by

$$W = \frac{hc}{\lambda} \quad (J) \tag{2.1}$$

where

h = Planck's constant (6.63×10^{-34} J s)
c = velocity of light (3×10^8 ms^{-1})
λ = wavelength of the emitted radiation

This equation pinpoints one potential difficulty of infrared thermography: wavelengths are large (with respect to visible spectrum wavelengths, for instance), consequently the photonic emitted energy will be small and of the same order of magnitude as radiation emitted by the room-temperature environment. Figure 2.1 illustrates these perturbing undesirable sources of energy.

An infrared camera (at least those deployed for quantitative applications) is not

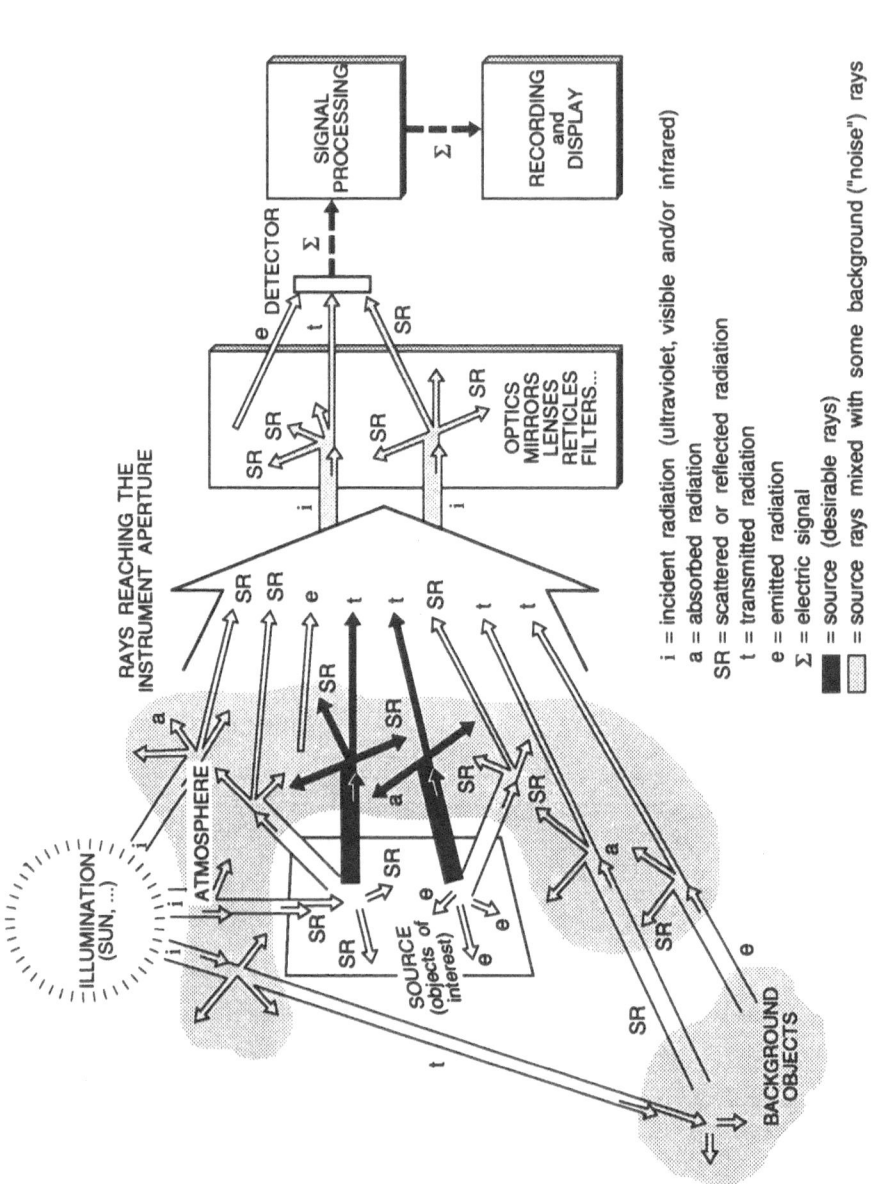

i = incident radiation (ultraviolet, visible and/or infrared)
a = absorbed radiation
SR = scattered or reflected radiation
t = transmitted radiation
e = emitted radiation
Σ = electric signal
■ = source (desirable rays)
▢ = source rays mixed with some background ("noise") rays

Fig. 2.1. Generalized radiometric configuration.

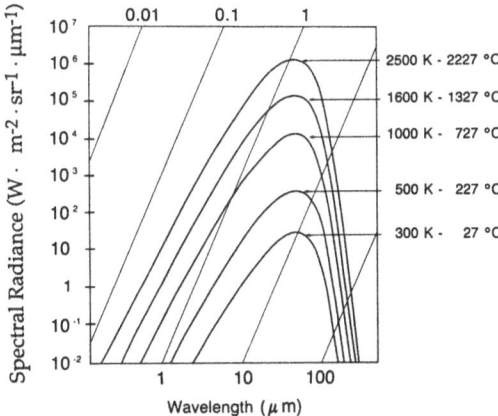

Fig. 2.2. Spectral radiance curves for a black body (Planck's law). Notice the locus of maximum values is located on the same line following Wien's law, $\lambda_{max} = 2898\,T^{-1}(K)$.

a thermometer but a *radiometer*. If some hypotheses are respected and if knowledge of some parameters is available, it thus becomes possible to translate radiometric values registered by the camera into temperature values.

This brings us to define the radiation emitted by the *black body*. The black body serves as a reference for the thermal emission of solids. An ideal black body is capable of absorbing totally all incident radiation, whatever the wavelength. In the same way, it also re-emits this radiation uniformly in all directions. If the thermal equilibrium is reached, for a black body at temperature $T(K)$, the spectral radiance $N_{\lambda,b}$ is given by the Planck's law (Fig. 2.2):

$$N_{\lambda,b} = \frac{2hc^2}{\lambda^5}\,\frac{1}{\exp\left(\dfrac{hc}{\lambda K T} - 1\right)} \qquad (\mathrm{W\,m^{-2}\,sr^{-1}\,\mu m^{-1}}) \qquad (2.2)$$

where h is Planck's constant, c is the velocity of light, k is Boltzmann's constant $(1.381 \times 10^{-23}\,\mathrm{JK^{-1}})$, λ is the wavelength of the emitted radiation (μm) and T is the temperature of the black-body cavity (K) and subscript b denotes black body. Black-body behaviour exists in the case of closed enclosures with opaque walls. Radiation is then emitted by a small orifice in order not to perturb the thermal equilibrium.

In the case of real objects whose absorbance is limited, only part of the energy will be radiated out from the surface, this fraction of the black-body spectral radiance is given by the property of the surface called the emissivity ε. For instance an object with emissivity of 0.5 will emit only half the total energy radiated by a black body at the same temperature. Objects whose emissivity is independent of the wavelength are called *grey bodies*. For objects called *coloured bodies*, emissivity depends upon the wavelength, the orientation and the temperature. We can write the expression of the radiance for such objects:

$$N_\lambda = \varepsilon N_{\lambda,b}(\lambda, T) \qquad (\mathrm{W\,m^{-2}\,sr^{-1}\,\mu m^{-1}}) \qquad (2.3)$$

More generally, emissivity $\varepsilon(\lambda, T, \theta)$ is a function also of the viewing angle θ (see for instance Fig. 11.4), of the temperature T and of the wavelength λ. In the case of metals for instance, emissivity increases with temperature (and is inversely proportional to the electrical conductivity (Gaussorgues 1984b, p 52).

For opaque objects, Kirchhoff laws relate the emissivity ε of a surface patch to the reflectivity ρ. For an incident isotropic radiation, we can write

$$\varepsilon = 1 - \rho \tag{2.4}$$

Equations (2.3) and (2.4) highlight one of the main problems of infrared thermography we discussed in Chap. 1. A surface having a low emissivity emits weakly (Eq. (2.3)) and has a high reflectivity coefficient ρ (Eq. (2.4)). Consequently, it tends to reflect incident radiation directly to the detector.

From these considerations, we can introduce the fundamental equation of thermography. The radiance N_{CAM} received by the camera is expressed by

$$N_{CAM} = \underset{\substack{\text{Object} \\ \text{contribution}}}{\tau_{atm}\varepsilon N_{obj}} + \underset{\substack{\text{Surrounding} \\ \text{environment} \\ \text{contribution}}}{\tau_{atm}(1 - \varepsilon)N_{env}} + \underset{\substack{\text{Atmosphere} \\ \text{contribution}}}{(1 - \tau_{atm})N_{atm}} \tag{2.5}$$

where

τ_{atm} = transmission coefficient of the atmosphere in the spectral window of interest

ε = object emissivity (the object is considered opaque)

N_{obj} = radiance from the surface of the object

N_{env} = radiance of the surrounding environment considered as a black body

N_{atm} = radiance of the atmosphere, supposed constant

Equation (2.5) can be simplified if we consider the transmission coefficient of the atmosphere as being close to unity. This hypothesis is justified for standard spectral bandwidths of operation (3 to 5 μm and 8 to 12 μm), over small distances (under 2 m) and if no absorbing gas, dust or water vapour droplet is present,

$$N_{CAM} = \varepsilon N_{obj} + (1 - \varepsilon)N_{env} \tag{2.6}$$

Relation (2.6) can be further reduced if the emissivity ε is high and if no high temperature object is present close to the inspection station which would give rise to parasitic reflections. High emissivity can be obtained if the object surface is covered with a special paint of high emissivity such as the Tremclad™ flat black or 3 M Black Velvet™ for which $\varepsilon \approx 0.9$ (see Chap. 7 for other considerations concerning the emissivity). If black-painting of the surfaces is not practical, techniques of thermal-transfer imaging can be used profitably (Maldague et al. 1991a). In Chaps 4, 5 and 6 we will see how relative measurements help to reduce emissivity problems. Under these conditions:

$$N_{obj} \approx N_{CAM}$$
$$f(N_{obj}) \approx I_{CAM} \tag{2.7}$$
$$S(T_{obj}) \approx I_{CAM}$$

where I_{CAM} corresponds to the radiometric signal obtained on the camera calibration curve (see Fig. 2.3); $f(\cdots)$ and $S(\cdots)$ are the relationships which allow us to convert camera signals to radiance N_{obj} and to temperature T_{obj} values respectively.

In the case of an ideal instrument a direct relationship could be derived between the radiometric signal I_{CAM} and the temperature of the object T_{obj} taking into account the limitations of Eq. (2.7). This ideal situation does not exist in general and we must consider the intrinsic limitations of the camera to perform truly quantitative measurements. These limitations are various:

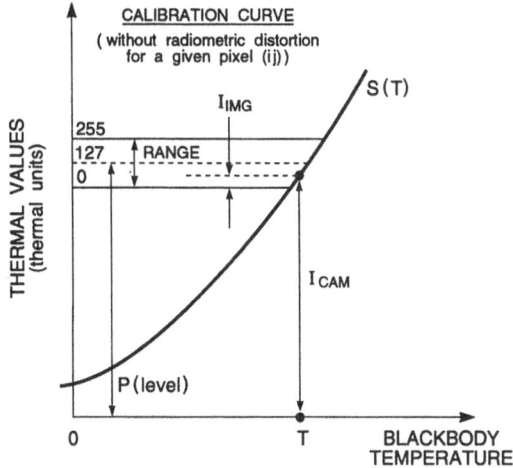

Fig. 2.3. Infrared camera calibration curve for temperature computation.

1. The vignetting and aberration of the optical system
2. The nonuniform spectral response of the instrument
3. The nonlinear response of the instrument in its dynamic range
4. The mechanical scanning process used for image formation which, if present, introduces spatial distortions in the images (Sect. 3.1.1 elaborates more on the image formation process)
5. The sensitivity of the infrared detector within the camera to self-emission by the camera (uncooled optical components such as lenses, mirrors and prisms are at ambient temperature and emit accordingly)
6. The Narcissus effect, by which the detector sees itself because of its own emission reflected in the optics

For instance, if transmittance changes 5% through the field of view due to vignetting and if a temperature difference of 20 K exists between the scene and the detector, a shadow effect of about 1 K will be observable (Abel 1977).

Following a proper calibration procedure, as we will see later (in Sect. 3.2), these uncertainties can be strongly reduced. The signal $I_{img}(i, j)$ is the digitized value (on 8 bits for instance, Fig. 2.3) actually obtained for the pixel at location (i, j) in the thermogram ("pixel" is derived from Picture Element). $I_{img}(i, j)$ is obtained from the camera pointed to the object surface patch of temperature T corresponding to position (i, j), (row, column) of the image; it can be related to $T(i, j)$ by (for 8 bit systems):

$$S[T(i, j)] = R(i, j)\left(\frac{I_{img}(i, j) - 128}{256}\right)\text{range} + P(\text{level}) + A(i, j) \qquad (2.8)$$

The first term of (2.8) corresponds to the effects of distortions due to differences in the optical path and vignetting, especially if expansion rings are mounted between the objective and the camera case to limit the field of view and obtain a close-up; these effects are corrected with $R(i, j)$.

The second term of (2.8) concerns adjustments of the dynamic range of the camera with respect to the temperatures being observed in the scene. Most infrared radiometers have two thermal adjustments, a *range* knob and a *level* knob. These knobs are set in order to avoid thermal image saturations of the objects of interest within the scene. The dynamic *range* is the maximum observable temperature differential (i.e. the non-saturated observable temperature span); it is adjusted to a given position along the calibration curve by the user. The level specifies the position of this maximum observable temperature differential on the calibration curve (Fig. 2.3). Moreover, the level is also the electric signal available which corresponds to the position of the level knob on the infrared camera front panel; this signal can be digitized (on 12 bits, 0–4095, for instance). More realistically, it is better to consider a function of the level, $P(\text{level})$.

The maximum resolution is obtained for the smallest range. Since we are interested to obtain the maximum resolution from the system, it is better to always use the smallest range available, if possible. The system is calibrated for one range value at a time. Other calibration sessions can be done for other range settings, if necessary. To recover temperature quantitatively it is thus necessary to record both the pixel values within the image and the level signal.

The third term of (2.8) concerns the correction needed for the self-emission effects of the camera. In an ideal instrument, neglecting the distortions, for a given pixel (i,j) in the image, we would have (from Fig. 2.3)

$$S[T(i,j)] = I_{\text{CAM}}(i,j) = \left(\frac{I_{\text{img}}(i,j) - 128}{256}\right)\text{range} + P(\text{level}), \qquad (2.9)$$

with

$$R(i,j) = 1 \quad \text{and} \quad A(i,j) = 0 \qquad \forall i, j \qquad (2.10)$$

for all (i, j) pixels and Eq. (2.8) would simplify to Eq. (2.9). For a nonideal instrument, matrices A and R are obtained after proper calibration (Sect. 3.1). We must point out that in Eq. (2.8) the level signal $P(\text{level})$ is considered uniform through the image plane even if practically, it can be modified at any time (during the image scanning) by the camera operator. However, the use of this relation allows us to perform an adequate radiometric correction process, as we will see later in Chap. 3.

Temperature can be computed if we consider the function $S(\cdots)$ relating the object surface of temperature T and the pixel intensity I_{img} at position (i, j) in the image. For temperature computations, we can rewrite (2.8) thus:

$$T(i,j) = S^{-1}\left\{R(i,j)\left(\frac{I_{\text{img}}(i,j) - 128}{256}\right)\text{range} + P(\text{level}) + A(i,j)\right\} \qquad (2.11)$$

We will see in Chap. 3 now this expression can be used to calibrate the infrared camera.

All of what we have seen up to now allows us to relate the temperature of the observed surface to the video signal obtained from the infrared camera, taking into account radiometric and geometric distortions. Signal degradation will be studied further in Sect. 3.2. In the next section, we examine the link between surface temperature and the presence of possible subsurface defect(s) taking into account the thermal stimulation scheme discussed in Chap. 1.

2.2 Heat Transfer Modelling

2.2.1 Analytical Approach

In order to establish a link between abnormal isotherms recorded on the inspected workpiece surface following the application of an external thermal stimulation (Sect. 1.2.2) and the presence of a subsurface defect, it is necessary to study thermal front propagation inside the material. As mentioned in Chap. 1, this is a very complex study if all the variables must be taken into consideration. The complexity of the analytical approach takes tremendous proportions as soon as we do not consider the ideal case of two flat surfaces of a uniform and isotropic material (i.e. a material having the same physical properties in all directions) and neglecting moreover the heat transfers inside both the defect itself and outer sample surfaces (Siegel and Howell 1972).

In a solid body, these transfers are governed by the Fourier diffusion equation (Arpaci, 1966):

$$\frac{\partial T}{\partial t} = \frac{K}{\rho C_p}\left(\frac{\partial^2 T}{\partial x^2} + \frac{\partial^2 T}{\partial y^2} + \frac{\partial^2 T}{\partial z^2}\right) \tag{2.12}$$

where T is the temperature, t the time, ρ the density, K the thermal conductivity and C_p the specific heat. Table 1.1 lists these thermo-physical properties for some commonly used materials.

To illustrate the analytical approach, we will briefly study a simple example (adapted from the work of Williams et al. (1980) and Sayers (1984). Equation (2.12) can be simplified and analytically solved if we consider the case of a semi-infinite isotropic plate. This plate spans infinitely in the space along x and y axes, but has a finite thickness along the z axis ($0 \leqslant z \leqslant l$). This corresponds to a one-dimensional case (Lau et al. 1990). Both plate surfaces are considered perfectly thermally insulated (no thermal loss), but the side on $z = 0$ is heated with an instantaneous and uniform thermal pulse. For the sake of simplicity, we consider that at time $t = 0$, the plate temperature is zero (no temperature differential with respect to the environment) but in a narrow region ($0 \leqslant z \leqslant z_e$, $z_e \ll l$) the temperature is T_0. For $t > 0$, the heat from this narrow region diffuses in the bulk of the plate. Of course, for very long times ($t \to \infty$), the temperature distribution becomes uniform with a final value of $T_f = T_0 z_e/l$.

Taking this geometry into consideration, Eq. (2.12) can be reduced to its unidimensional equivalent with $T(z,t)$ corresponding to temperature distribution at time t and at depth z:

$$\frac{\partial^2}{\partial z^2} T(z,t) = \frac{\rho C}{K}\frac{\partial T(z,t)}{\partial t} \tag{2.13}$$

The boundary conditions are

$$\begin{aligned} T(z,0) &= 0 \quad z_e \leqslant z \leqslant l \\ T(z,0) &= T_0 \quad 0 \leqslant z \leqslant z_e \\ \frac{\partial T(0,t)}{\partial z} &= \frac{\partial T(l,t)}{\partial z} = 0 \end{aligned} \tag{2.14}$$

This problem has an analytical solution (Carslaw and Jaeger 1959):

$$T(z,t) = T_f\left[1 + \frac{2l}{\pi z_e}\sum_{n=1}^{\infty}\frac{1}{n}\sin\left(\frac{n\pi z_e}{l}\right)\cos\left(\frac{n\pi z}{l}\right)\exp\left(\frac{-t\alpha^2\pi^2 n^2}{l^2}\right)\right] \tag{2.15}$$

Analytical solutions (such as that of Eq. (2.15)) are very useful to check results obtained by numerical computation, as we will see in Sect. 1.2.2.

This example shows that even in the case of simple geometries, the analytical solution is relatively complex. The analytical approach rapidly degenerates into unworkable situations with the introduction of both anisotropic properties and subsurface defects, especially if one-dimensional approximation cannot be retained.

2.2.2 Finite Difference Modelling

For studies of more complex shapes, finite difference modelling is a precious tool, especially since it can provide limits of the effectiveness of the TNDE technique and also the possibility to consider different defect geometries and determine their detectability without the expense of making and testing the corresponding specimens (Burch et al. 1984; Baughn and Johnson 1986; James et al. 1989). This is called the direct problem. In this section we will present the basis of a simple "workable" model whose programming translation is given in C language at the end of this book (App. A). Although less powerful than some commercially available heat transfer models (such as I/TAS from Warriner Ass., 22 Margate Square, Palos Verdes, CA 90274, USA), the one we introduce in this section has the advantage of being easily understood and adaptable.

In a first step the workpiece to be studied is represented by a mesh of elementary volumes of material. A node is associated with each volume and is representative of the corresponding thermal behaviour of that volume. More specifically, the temperature of a node is considered as being equal to the average temperature of the surrounding material within the corresponding elementary volume.

A practical approach to laying out the mesh is to adopt a cylindrical system of coordinates (Fig. 2.4). Using this scheme, it is only necessary to compute results on one pie slice (Fig. 2.5) and because of the symmetry of this configuration, the same values can be applied all around ($0 < \theta < 2\pi$). Computation will proceed much more rapidly than for a *true* Cartesian geometry.

Each elementary volume (around a node) can be thought as a box having radial widths Δr and axial depth Δz. There are (nz) nodes in depth and (nr) nodes along the radial length.

The basic modelling idea is to compute the temperature for all the nodes, taking

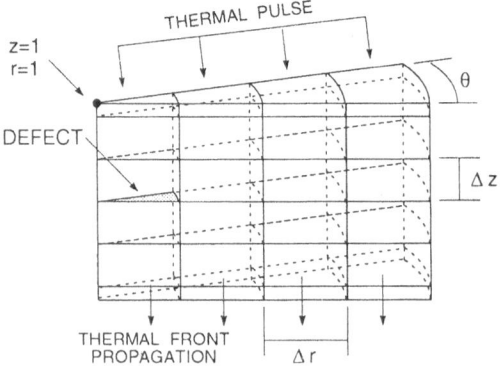

Fig. 2.4. Cylindrical reference system used in heat transfer modelling.

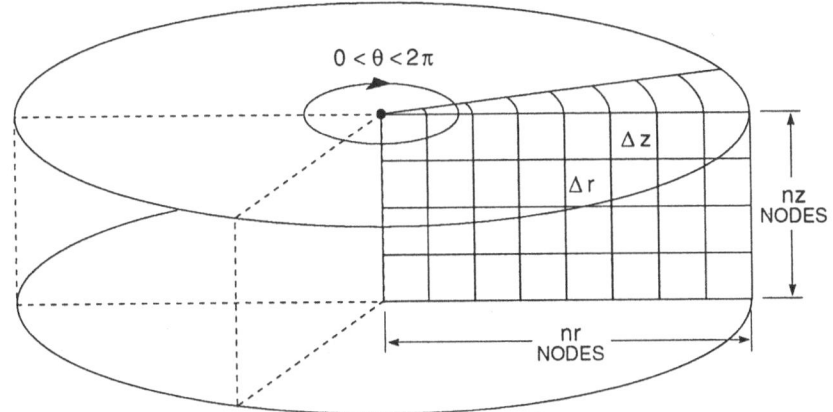

$$0 < \theta < 2\pi$$

Δz

Δr

nz
NODES

nr
NODES

NOTE : The grid represents the computed
area for this cylindrical geometry

Fig. 2.5. Cylindrical reference system used in heat transfer modelling. Because of the symmetry effect, it is only necessary to compute temperature distribution for the nodes shown on the right (matrix $nr \times nz$).

into account the thermal properties of the material and the thermal exchanges between the nodes themselves and the external world. As noted with Eq. (2.12), time is an important variable since we have a dynamic system. In active thermography, transient study after application of the thermal perturbation (at $t = 0$) is fundamental, as seen in Chap. 1, in order to provoke a response from the inspected part. In fact the permanent regime does not bring useful information about subsurface defects. In this respect, the time scale is also divided in small increments Δt and, for each time increment, that is for every iteration of the program, the temperature of all the nodes is re-computed.

This procedure allows sufficient accuracy of modelling to provide good insight into thermal phenomena, although it does not consider all the aspects. Notice for instance the thermal capacitance between nodes is ignored in this study as the resistive behaviour is prominent (Krapez et al. 1991).

Temperature computations (e.g. Thomas 1980) at all the nodes are done following the *first law of thermodynamics* (energy conservation), the *Fourier law of conduction* and the external thermal exchanges (energy losses and deposits). For a given instant t:

$$\text{Rate of energy creation} = 0$$

$$-\left[\sum \frac{\Delta E_{\text{out}}}{\Delta t} - \sum \frac{\Delta E_{\text{in}}}{\Delta t}\right] = \frac{\Delta E_{\text{s}}}{\Delta t} \qquad (2.16)$$

where E_{out} and E_{in} represent energy transferred into and out of a particular node and ΔE_{s} corresponds to the energy stored in the elementary volume surrounding the node. The minus sign means the flux exchanged is positive for a negative temperature gradient. This is consistent with the *second law of thermodynamics* (energy is transferred in the direction of decreasing temperature).

For our geometry (Figs. 2.4 and 2.5) we have both radial (Δq_r flux along radial direction r) and axial (Δq_z flux along the depth) transfers. Equation (2.16) can be

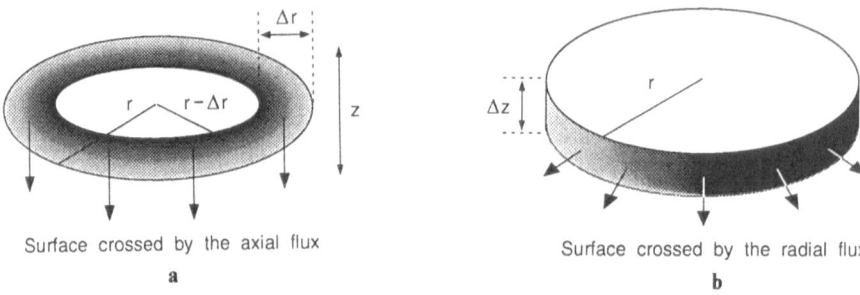

Fig. 2.6. Surfaces crossed by thermal flux: (a) axial; (b) radial.

rewritten, for the node located at (r, z):

$$-(\Delta q_r + \Delta q_z) = \frac{\Delta E_s}{\Delta t} = \frac{\rho \Delta V C_p \Delta T}{\Delta t} \qquad (2.17)$$

and thus

$$T_{future} = T_{present} - \frac{(\Delta q_r + \Delta q_z)\Delta t}{\rho \Delta V C_p} \qquad (2.18)$$

where ρ is the density, C_p is the specific heat at node (r, z) and ΔV corresponds to the elementary volume through which thermal exchanges occur, for the node in (r, z):

$$\Delta V = \Delta z(S_{crossed,z}) \qquad (2.19)$$

where $S_{crossed,z}$ is the surface crossed by the thermal flux from z to $(z + \Delta z)$ (Fig. 2.6a):

$$S_{crossed,z} = \pi(r^2 - (r - \Delta r)^2) \qquad (2.20)$$

Flux values Δq_r and Δq_z are obtained from the *Fourier law of conduction*, for the node in (r, z):

$$\Delta q_r = \frac{K_r S_{crossed,r}}{\Delta r} (T_{present\ in\ r} - T_{present\ in\ r+\Delta r}) \qquad (2.21)$$

where $T_{present\ in\ r}$ and $T_{present\ in\ r+\Delta r}$ are the temperatures at the present time for the adjacent nodes r and $r + \Delta r$ (radial direction), K_r is the radial thermal conductivity and $S_{crossed,r}$ is the surface crossed by the thermal front from r to $r + \Delta r$, from Fig. 2.6b:

$$S_{crossed,r} = 2\pi r \Delta z \qquad (2.22)$$

Similarly, for Δq_r, for the node located in (r, z):

$$\Delta q_z = \frac{K_z S_{crossed,z}}{\Delta z} (T_{present\ in\ z} - T_{present\ in\ z+\Delta z}) \qquad (2.23)$$

where $T_{present\ in\ z}$ and $T_{present\ in\ z+\Delta z}$ are the temperatures at the present time for the two adjacent nodes z and $z + \Delta z$ (axial direction), K_z is the axial thermal conductivity and $S_{crossed,z}$ is the surface crossed by the thermal front from z to $z + \Delta z$ (Eq. (2.20)).

All these computations (Eqs (2.18), (2.21) and (2.23)) must be performed along all thermal exchange directions for the nodes. For one *slice* of our cylindrical system of coordinates, there are four directions for a given node located in (r, z): $(r - \Delta r, z)$, $(r, z - \Delta z)$, $(r + \Delta r, z)$, $(r, z + \Delta z)$. The other two directions of space are taken into account by the cylindrical geometry: the temperature distribution is the same for all the slices. These effects are considered in the computation of the surfaces crossed by the thermal fluxes (Eqs (2.20) and (2.22)).

Considering now the initial thermal stimulation, a quantity of energy P (in $W\,cm^{-2}$) is deposited during a certain number of time steps ($0 < t <$ duration of the initial thermal stimulation) on the front surface (nodes in $z = 0$). From the surface, the thermal front propagates, by diffusion, inside the whole mesh. The increase of temperature ΔT_p on the front surface ($z = 0$) corresponding to this stimulation is given by

$$\Delta T_p = \frac{P\Delta t}{\rho C_p S_{crossed,z} \Delta z} \tag{2.24}$$

with the parameters defined as previously for Eq. (2.18).

Loss effects are very important to consider (Tossel 1989). In our model, losses (by radiation and convection) are only considered on the external face of the sample (nodes with $z = 0$). The opposite surface ($z = nz$) is supposed to be thermally well insulated. This hypothesis is close to the case of an *infinite* thickness sample. This is also the case of a sample of finite thickness but when the time of observation (following the initial thermal perturbation) is small with respect to the heat front propagation.

Radiative losses are computed after the *Stefan–Boltzmann law*, for the node located at $(x, y, z = 0)$:

$$\Delta q_{rad} = F_{rad}(T_s^4 - T_a^4) \tag{2.25}$$

where T_s is the temperature expressed in kelvins for the node on the front surface $(x, y, z = 0)$, T_a is the ambient temperature and F_{rad} is an empirical factor which depends on the geometry (see Table 2.1) (Sparrow and Chess 1978).

Convective losses are computed after the *Newton law of cooling*:

$$\Delta q_{con} = F_{con}(T_s - T_a) \tag{2.26}$$

with T_s and T_a defined as before and where F_{con} is an empirical factor which depends on the geometry (see Table 2.1).

These losses contribute to reduce the front temperature, for the node located in $(r, z = 0)$:

$$T_{present\ with\ losses} = T_{present\ no\ loss} - \frac{(\Delta q_{rad} + \Delta q_{con})\Delta t}{\rho C_p \Delta V} \tag{2.27}$$

Table 2.1. Modelling parameters

Δt	Variable logarithmic progression, 1200 iterations (30 s)
F_{rad}	$5.67 \times 10^{-12}\,W\,cm^{-2}$ (Cielo 1983)
F_{con}	$1 \times 10^{-3}\,W\,cm^2$ (Cielo 1983)
T_a	$0\,°C^a$
Δr	$2\,mm$
Δz	$200\,\mu m$
nr	20
nz	30
Heating duration	$2\,s$
P	$1\,W\,cm^{-2}$

[a] Equivalent to computing an increase of temperature.

Time is: 0.50 sec   Delta T max (for row 1, defect - sound area):   0.0 °C

4.224	4.224	4.224	4.224	4.224	4.224	4.224	4.224	4.224	4.224	4.224	4.224	4.224
3.286	3.286	3.286	3.286	3.286	3.286	3.286	3.286	3.286	3.286	3.286	3.286	3.286
2.120	2.120	2.120	2.120	2.120	2.120	2.120	2.120	2.120	2.120	2.120	2.120	2.120
1.296	1.296	1.296	1.296	1.296	1.296	1.296	1.296	1.296	1.296	1.296	1.296	1.296
.748	.748	.748	.748	.748	.748	.748	.748	.748	.748	.748	.748	.748
.406	.406	.406	.406	.406	.406	.406	.406	.406	.406	.406	.406	.406
.206	.206	.206	.206	.206	.206	.206	.206	.206	.206	.206	.206	.206
.097	.097	.097	.097	.097	.097	.097	.097	.097	.097	.097	.097	.097
.043	.043	.043	.043	.043	.043	.043	.043	.043	.043	.043	.043	.043
.017	.017	.017	.017	.017	.017	.017	.017	.017	.017	.017	.017	.017
.007	.007	.007	.007	.007	.007	.006	.006	.006	.006	.006	.006	.006
.003	.003	.003	.003	.003	.002	.002	.002	.002	.002	.002	.002	.002
.003	.000	.000	.000	.001	.001	.001	.001	.001	.001	.001	.001	.001
.000	.000	.000	.000	.000	.000	.000	.000	.000	.000	.000	.000	.000
.000	.000	.000	.000	.000	.000	.000	.000	.000	.000	.000	.000	.000
.000	.000	.000	.000	.000	.000	.000	.000	.000	.000	.000	.000	.000
.000	.000	.000	.000	.000	.000	.000	.000	.000	.000	.000	.000	.000
.000	.000	.000	.000	.000	.000	.000	.000	.000	.000	.000	.000	.000
.000	.000	.000	.000	.000	.000	.000	.000	.000	.000	.000	.000	.000
.000	.000	.000	.000	.000	.000	.000	.000	.000	.000	.000	.000	.000

Time is: 2.00 sec   Delta T max (for row 1, defect - sound area):   0.004705 °C

9.669	9.669	9.667	9.665	9.664	9.664	9.664	9.664	9.664	9.664	9.664	9.664	9.664
8.542	8.542	8.541	8.540	8.537	8.536	8.536	8.536	8.536	8.536	8.536	8.536	8.536
7.006	7.006	7.005	7.003	6.998	6.997	6.997	6.997	6.997	6.997	6.997	6.997	6.997
5.678	5.678	5.677	5.674	5.666	5.664	5.663	5.663	5.663	5.663	5.663	5.663	5.663
4.548	4.548	4.547	4.542	4.529	4.526	4.525	4.525	4.525	4.525	4.525	4.525	4.525
3.604	3.604	3.602	3.594	3.574	3.569	3.568	3.568	3.568	3.568	3.568	3.568	3.568
2.831	2.831	2.830	2.828	2.816	2.783	2.775	2.774	2.774	2.774	2.774	2.774	2.774
2.213	2.213	2.213	2.212	2.209	2.192	2.138	2.128	2.127	2.126	2.126	2.126	2.126
1.737	1.737	1.736	1.731	1.707	1.621	1.608	1.606	1.606	1.606	1.606	1.606	1.606
1.387	1.387	1.386	1.381	1.349	1.213	1.197	1.195	1.195	1.195	1.195	1.195	1.195
1.155	1.155	1.154	1.153	1.147	1.107	.893	.877	.876	.876	.876	.876	.876
1.029	1.029	1.028	1.021	.975	.642	.632	.632	.632	.632	.632	.632	.632
.051	.051	.052	.059	.105	.438	.448	.448	.448	.448	.448	.448	.448
.034	.035	.036	.042	.082	.312	.313	.313	.313	.313	.313	.313	.313
.023	.023	.024	.029	.061	.198	.213	.215	.215	.215	.215	.215	.215
.015	.015	.016	.020	.044	.130	.143	.145	.145	.145	.145	.145	.145
.010	.010	.010	.014	.031	.085	.095	.096	.096	.096	.096	.096	.096
.006	.006	.007	.009	.021	.054	.062	.063	.063	.063	.063	.063	.063
.004	.004	.004	.006	.014	.034	.039	.040	.040	.040	.040	.040	.040
.002	.002	.003	.004	.009	.021	.025	.025	.025	.025	.025	.025	.025

Time is: 10.78 sec Delta T'max (for row 1, defect - sound area): 0.859981 °C

3.280	3.235	3.144	3.007	2.838	2.670	2.554	2.487	2.452	2.434	2.426	2.423	2.421	2.421	2.420	2.420	2.420	2.420	2.420	2.420
3.284	3.239	3.147	3.009	2.837	2.667	2.550	2.483	2.448	2.431	2.423	2.420	2.418	2.418	2.417	2.417	2.417	2.417	2.417	2.417
3.282	3.237	3.144	3.005	2.829	2.652	2.535	2.468	2.434	2.417	2.410	2.406	2.405	2.404	2.404	2.404	2.404	2.404	2.404	2.404
3.275	3.229	3.136	2.994	2.812	2.627	2.508	2.443	2.410	2.394	2.386	2.383	2.382	2.381	2.381	2.381	2.381	2.381	2.381	2.381
3.262	3.216	3.122	2.977	2.788	2.590	2.470	2.407	2.375	2.360	2.353	2.350	2.349	2.348	2.348	2.348	2.348	2.348	2.348	2.348
3.243	3.197	3.103	2.955	2.756	2.541	2.421	2.361	2.332	2.317	2.311	2.308	2.307	2.306	2.306	2.306	2.306	2.306	2.306	2.306
3.219	3.174	3.078	2.928	2.719	2.481	2.362	2.306	2.279	2.266	2.260	2.258	2.256	2.256	2.256	2.256	2.256	2.256	2.256	2.256
3.190	3.145	3.050	2.896	2.677	2.409	2.293	2.242	2.218	2.207	2.202	2.199	2.198	2.198	2.198	2.198	2.198	2.198	2.198	2.198
3.156	3.112	3.017	2.862	2.631	2.324	2.215	2.171	2.150	2.141	2.136	2.134	2.133	2.133	2.133	2.133	2.133	2.133	2.133	2.133
3.118	3.074	2.980	2.825	2.584	2.224	2.128	2.092	2.076	2.068	2.064	2.063	2.062	2.062	2.062	2.062	2.062	2.062	2.062	2.062
3.074	3.032	2.940	2.786	2.538	2.109	2.034	2.008	1.996	1.990	1.988	1.986	1.986	1.985	1.985	1.985	1.985	1.985	1.985	1.985
3.027	2.985	2.897	2.747	2.499	1.973	1.934	1.920	1.912	1.908	1.906	1.906	1.905	1.905	1.905	1.905	1.905	1.905	1.905	1.905
.869	.901	.971	1.096	1.316	1.813	1.830	1.828	1.825	1.823	1.822	1.822	1.821	1.821	1.821	1.821	1.821	1.821	1.821	1.821
.818	.852	.925	1.053	1.272	1.673	1.725	1.734	1.736	1.736	1.736	1.736	1.736	1.736	1.736	1.736	1.736	1.736	1.736	1.736
.769	.803	.878	1.007	1.218	1.548	1.620	1.640	1.646	1.648	1.648	1.649	1.649	1.649	1.649	1.649	1.649	1.649	1.649	1.649
.722	.756	.831	.959	1.158	1.434	1.519	1.546	1.556	1.559	1.561	1.561	1.562	1.562	1.562	1.562	1.562	1.562	1.562	1.562
.676	.711	.785	.909	1.096	1.331	1.420	1.454	1.467	1.472	1.474	1.475	1.475	1.476	1.476	1.476	1.476	1.476	1.476	1.476
.633	.667	.740	.860	1.032	1.235	1.326	1.364	1.380	1.387	1.389	1.391	1.391	1.391	1.391	1.391	1.391	1.391	1.391	1.391
.591	.625	.696	.810	.970	1.147	1.237	1.278	1.296	1.304	1.308	1.309	1.310	1.310	1.310	1.310	1.310	1.310	1.310	1.310
.552	.585	.654	.763	.909	1.066	1.154	1.197	1.216	1.225	1.229	1.231	1.232	1.232	1.232	1.232	1.232	1.232	1.232	1.232

Time is: 20.0 sec Delta T max (for row 1, defect - sound area): 0.559669 °C

2.334	2.302	2.241	2.155	2.056	1.960	1.889	1.844	1.815	1.798	1.788	1.782	1.778	1.776	1.775	1.775	1.774	1.774	1.774	1.774
2.338	2.306	2.245	2.159	2.058	1.961	1.891	1.845	1.817	1.800	1.790	1.784	1.781	1.779	1.778	1.777	1.777	1.777	1.777	1.777
2.339	2.307	2.246	2.159	2.058	1.959	1.888	1.844	1.816	1.799	1.790	1.784	1.780	1.778	1.777	1.777	1.777	1.776	1.776	1.776
2.338	2.306	2.245	2.157	2.054	1.952	1.882	1.838	1.812	1.796	1.786	1.781	1.777	1.775	1.774	1.774	1.774	1.773	1.773	1.773
2.334	2.302	2.241	2.153	2.047	1.941	1.872	1.830	1.804	1.789	1.780	1.775	1.772	1.770	1.769	1.768	1.768	1.768	1.768	1.768
2.327	2.296	2.234	2.146	2.037	1.926	1.858	1.818	1.794	1.780	1.771	1.766	1.763	1.762	1.761	1.760	1.760	1.760	1.760	1.760
2.318	2.286	2.225	2.136	2.025	1.906	1.840	1.803	1.781	1.768	1.760	1.756	1.753	1.751	1.751	1.750	1.750	1.750	1.750	1.750
2.306	2.275	2.214	2.125	2.010	1.881	1.818	1.785	1.765	1.754	1.747	1.743	1.740	1.739	1.738	1.738	1.737	1.737	1.737	1.737
2.291	2.261	2.201	2.112	1.994	1.851	1.793	1.764	1.747	1.737	1.731	1.727	1.725	1.724	1.723	1.723	1.723	1.723	1.723	1.723
2.274	2.245	2.186	2.098	1.977	1.816	1.765	1.741	1.727	1.719	1.714	1.710	1.709	1.708	1.707	1.707	1.706	1.706	1.706	1.706
2.255	2.226	2.169	2.082	1.960	1.773	1.734	1.716	1.705	1.698	1.694	1.692	1.690	1.689	1.689	1.689	1.688	1.688	1.688	1.688
2.234	2.206	2.150	2.066	1.944	1.723	1.700	1.689	1.682	1.677	1.674	1.672	1.671	1.670	1.670	1.669	1.669	1.669	1.669	1.669
1.243	1.263	1.303	1.366	1.464	1.662	1.665	1.661	1.657	1.654	1.652	1.651	1.650	1.649	1.649	1.649	1.649	1.649	1.649	1.649
1.220	1.240	1.282	1.347	1.446	1.609	1.629	1.632	1.631	1.630	1.629	1.629	1.628	1.628	1.628	1.628	1.627	1.627	1.627	1.627
1.197	1.218	1.261	1.328	1.425	1.563	1.594	1.602	1.605	1.606	1.606	1.606	1.606	1.606	1.606	1.606	1.606	1.606	1.606	1.606
1.174	1.196	1.240	1.307	1.402	1.520	1.559	1.573	1.578	1.581	1.582	1.583	1.583	1.583	1.583	1.583	1.583	1.583	1.583	1.583
1.152	1.174	1.219	1.286	1.377	1.482	1.525	1.543	1.552	1.556	1.559	1.560	1.561	1.561	1.561	1.561	1.561	1.561	1.561	1.561
1.131	1.153	1.198	1.265	1.352	1.446	1.493	1.515	1.526	1.532	1.536	1.537	1.538	1.539	1.539	1.539	1.539	1.539	1.539	1.539
1.110	1.133	1.178	1.244	1.327	1.414	1.462	1.487	1.501	1.508	1.513	1.515	1.516	1.517	1.517	1.518	1.518	1.518	1.518	1.518
1.091	1.113	1.158	1.223	1.303	1.384	1.433	1.461	1.477	1.486	1.491	1.494	1.495	1.496	1.497	1.497	1.497	1.497	1.497	1.497

Fig. 2.7. Temperature distribution computed for a graphite epoxy specimen. The position of a thin air layer (subsurface flaw) is indicated by a line. The results are shown for four different times. (See text for more details.)

The parameters are defined as before (Eq. (2.18)).

This set of equations has been used to simulate various kinds of experimental configurations, as we will see in the later sections.

In Fig. 2.7, we show the temperature distribution computed using this model for different times (geometries of Figs 2.4 and 2.5). The temperature is given for the first 20 rows and the first 20 columns of the mesh in the case of a graphite epoxy sample, with the following model parameters: radial conductivity $K_r = 0.05$ W cm^{-1} °C^{-1}, axil conductivity $K_z = 0.01$ W cm^{-1} °C^{-1}, specific heat $C_p = 1000$ J Kg^{-1} °C^{-1}, mass density $\rho = 2000$ kg m^{-3}. Table 2.1 shows the values used for computation of Fig. 2.7.

We remember that in the case of our cylindrical geometry, the results in Fig. 2.7 correspond to one slice (Fig. 2.4). These results reveal some important facts:

Lateral diffusion of the thermal front around the defect is important (especially at $t = 10.78$ s). This *flow* effect reduces the visibility of the defect.

Temperature differentials are smaller on the surface than at the interface of the defect itself; this also reduces the visibility of the defect.

The surface temperature above the defect is not constant and reaches a maximum at a specific time (which depends on the defect depth, Eq. (1.1) and Fig. 1.2); consequently, there is an optimum time of observation. In the case of the geometry of Fig. 2.7, the maximum temperature differential is observed at $t = 10.78$ s.

As we will see in the following chapters, this kind of analysis is very useful to predict and understand experimental results.

The accuracy that can be obtained with such modelling is strongly related to the size of the elementary volumes considered (Δr, Δz, Δt). With very small values, errors introduced by the finite difference approximations reduce drastically at the expense of a tremendous increase in the computation time and risks of nonconvergence. Of course a well balanced compromise between the needed accuracy (depending on the infrared camera resolution if comparisons are to be made with experimental results, for instance), the size of the elementary volumes and the time step size must be made. In order to specify values for ($\Delta r, \Delta z, \Delta t$), we can start with large values and then solve the problem with smaller and smaller values of ($\Delta r, \Delta z, \Delta t$). Observing the convergence of the distribution of temperature allows the computation to be stopped when an acceptable accuracy is obtained.

It is also important to specify values respecting the stability criteria required for the convergence of the computation (Cielo 1984):

$$(\Delta z)^2 \geqslant 6\alpha\Delta t$$
$$(\Delta r)^2 \geqslant 6\alpha\Delta t$$

$$(2.28)$$

in order to prevent temporal instability which is likely to occur otherwise in the iterative computation procedure.

In order to validate such modelling, two directions can be adopted. The model can be applied to a very simple problem (for instance an infinite plate) for which an exact analytical solution exists (e.g. Eq. (2.15)). This does not mean, however, that the model will perform satisfactorily in complex situations. Probably a better approach is to compare the model with data acquired from a corresponding experimental set-up. Care must be taken when performing such comparisons since experimental values can be corrupted with adverse phenomena not taken into account in the model such as nonuniform surface thermal stimulation, uneven surface emissivity, complex loss phenomena, variable material properties, etc. Even if comparisons with experimental data are not straightforward, such modelling brings useful information about typical thermal behaviour for specific families of samples. It also helps to optimize TNDE parameters such as length and strength of thermal stimulation, and the expected optimum time window of observation for best defect visibility.

Conclusion. In this chapter, we have considered theoretical aspects of radiometry and heat transfer modelling which are fundamental in the deployment of the TNDE, as will be seen in the following chapters.

Chapter 3

Experimental Apparatus

3.1 Description of the System and Intended Use

A system used for thermographic inspection and based on principles presented in the previous chapters must be composed of many different subsystems (Spiro and Schlessinger 1989) such as the infrared (IR) camera, the image acquisition and analysis system and the thermal stimulation system (Fig. 3.1). In this chapter, we will focus our attention on the main elements of the thermographic inspection system illustrated in Fig. 3.1.

In this book we concentrate particularly on imaging systems which are of more interest for fast TNDE inspection procedures than point or line systems (cf. Chap. 1). Obviously this two-dimensional study can also be applied (partly) to line or point systems. In this section we will review the elements of a typical TNDE apparatus and in order to illustrate this study, we will refer to a real TNDE inspection station.

There are different ways in which infrared equipment can be deployed for TNDE depending on the specific application which is envisaged. As specified in Chap. 1, we can say than thermography can be deployed either in a *passive* or an *active* fashion. In the passive configuration the infrared imaging camera is pointed at the scene, looking at objects to be inspected, while no external thermal perturbation is applied. Applications such as building inspection, control of industrial processes, maintenance evaluation, power generation investigation, military surveys, welding, forest fire watching and medicine make use of this mode of operation where the surface temperature distribution, as is, contains relevant information concerning possible disorders, presence or absence of targets, etc.

In the active configuration an externally applied thermal stimulation is needed to generate meaningful contrasts which will yield to the detection of subsurface abnormalities (Chap. 1). Without this, no information can be drawn from the inspection since surface temperature distribution is not related to subsurface structure. Applications of active thermography include material analysis and parts inspection. In this chapter, we concentrate on deployment of infrared equipment in regard to the active configuration although, besides the thermal perturbation source used for part stimulation, the following discussion is also relevant to the passive configuration.

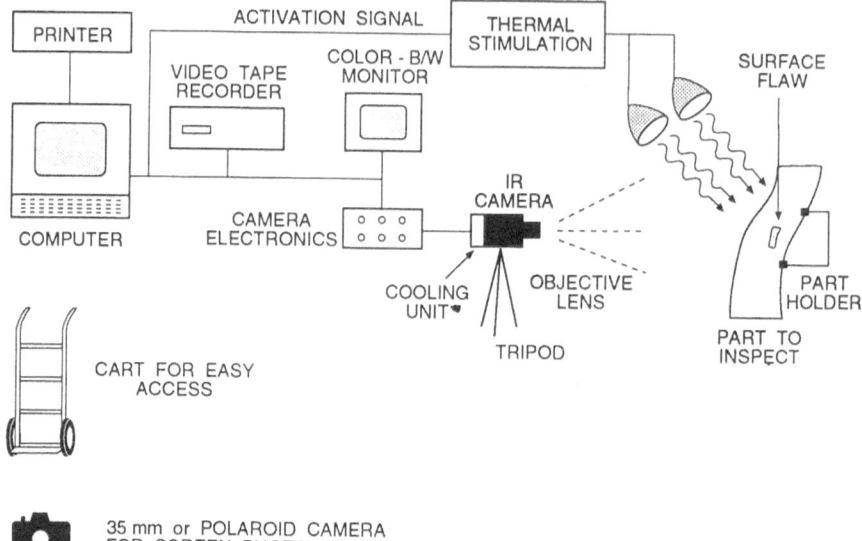

Fig. 3.1. Typical active infrared NDE experimental set-up.

3.1.1 Infrared Cameras and Infrared Images

Image Formation

When quantitative analysis is required, an infrared radiometer is necessary. The difference between such an instrument and an ordinary infrared camera is that, for the radiometer, the infrared signal is temperature calibrated thanks to the presence of internal temperature references seen by the detector elements during the image formation process. This calibration signal allows recovery to the absolute temperature after proper processing (Sect. 3.3). In this instrument, the image is electro-mechanically scanned over the detector surface by means of the synchronous rotation of mirrors or prisms (Fig. 3.2). To accommodate the standard video signal format of 30 frames per second (25 in Europe and part of Japan), a very high scanning rate is required. This imposes a wide bandwidth from the associated electronics for the noise level to be kept small. To overcome this problem, some manufacturers use slower scanning rates and have frames made of several fields with one field update at each scan (e.g. 4 : 1). As a result, the output obtained, comprising several fields, is not updated in real time (that is at 30 or 25 Hz). This may cause problems when fast thermal events must be observed since thermal information registered at different times is mixed together. One way to solve this problem is to split the video signal into its basic fields and process them, knowing the time interval between each (Shepard and Sass 1990; Shepard et al. 1991). However, this kind of analysis requires careful manipulation of the signal.

If a radiometer is not available, it is still possible to perform quantitative investigations with an infrared camera (e.g. based on a pyroelectric vidicon tube or on an FPA sensor, Sect. 1.4) at the condition to place a known temperature reference source

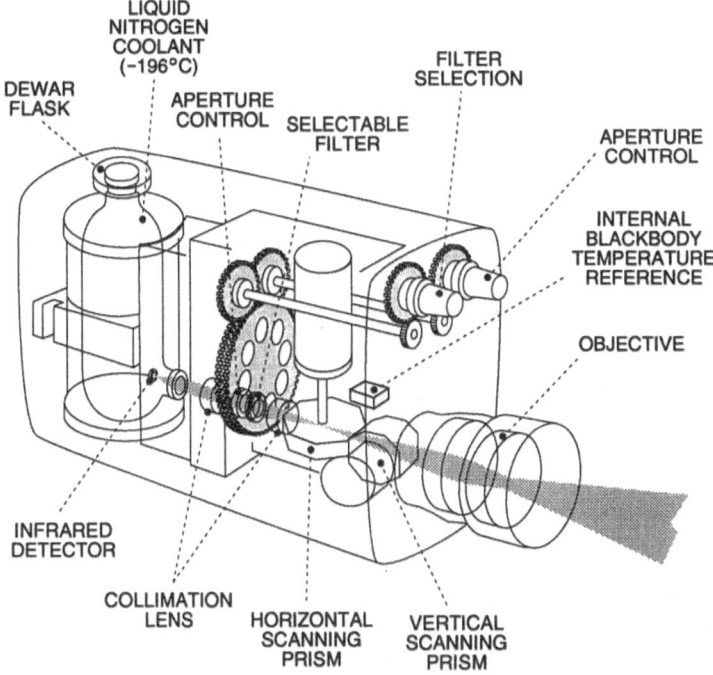

Fig. 3.2. Image formation: internal view of an infrared radiometer.

in the field of view. This may not always be possible for example, in the case of distant objects or restricted fields of view, moreover this reduces also the useful portion of the image.

Various lenses and expansion rings can be mounted in front of the IR camera in order to accommodate a specific field of view. Expansion rings or tubes inserted in the optical path between the objective and the camera permit achievement of greater magnification. Care should be taken, however, to be aware of the geometric distortions and vignetting effects which are likely to occur when high magnification is used. Proper calibration (Sect. 3.3) corrects for these effects.

Important Camera Characteristics

Thermal imagers can be characterized by many different parameters of interest (Williams and Davidson 1989; Reichenbach et al. 1991). One of the main characteristics of infrared detectors is the *normalized detectivity* denoted D^* (Levinstein 1965). This quantity is defined as the signal to noise ratio (SNR) per watt of incident power and is normalized both for an $1\,cm^2$ detector sensitive surface and $1\,Hz$ bandwidth. Figure 1.9 presents the spectral detectivity curves for the most common detectors. For instance, a pyroelectric tube has a $D^* \sim 10^8$ to $10^9\,cm\,Hz^{1/2}\,W^{-1}$. In the case of InSb (indium antimonite), $D^* \sim 6 \times 10^{10}\,cm\,Hz^{1/2}\,W^{-1}$ at an operation temperature of $-196\,°C$ (77 K).

Noise equivalent temperature difference (NETD) is another important specification of infrared cameras. For instance an NETD of 0.2 °C at 30 °C means that the noise level at the camera output is equivalent to a temperature differential of 0.2 °C on a black body heated at 30 °C. Another important figure of merit concerning infrared cameras (Wood et al. 1976; Holst and Pickard 1989; Gao et al. 1991) is the minimum resolvable temperature difference (MRTD) which is the minimum temperature difference observable by an operator on a target constituted by a periodic bar pattern. MRTD is a rather subjective measurement since the operator can adjust camera settings as wished in order to get the best MRTD value.

Another important characteristic of an imaging system (not only infrared) is the line spread function (or LSF) which allows evaluation of the spatial geometrical quality of a camera, that is its ability to reproduce correctly the spatial frequencies of the observed object (Wong 1982, Eq. 2.23, p. 23; Wood 1982; Holmsten 1986; Oermann 1987; Harding 1988; Reichenbach et al. 1991; Ryu 1991; Kennedy 1991). In fact the LSF specifies the spatial resolution of the detector; this characteristic determines how a sine pattern of given spatial frequency will be reproduced. Knowing the LSF, the raw image could be restored by dividing (if not zero) its Fourier transform by the detector's LSF expressed in the frequency domain:

$$\text{Recorded image} \approx \text{LSF} * \text{Original image} \qquad (3.1)$$

where $*$ stands for the convolution operator (convolution corresponds to multiplication in the Fourier frequency domain (Gonzalez and Wintz 1987, Chap. 3).

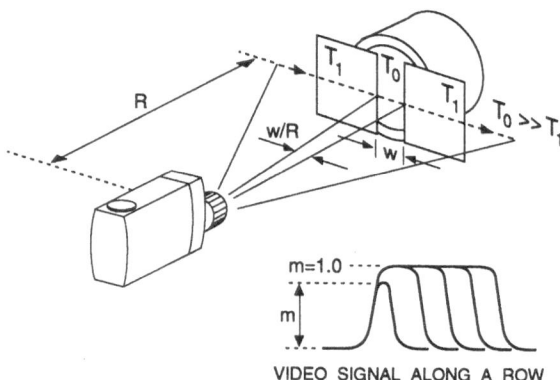

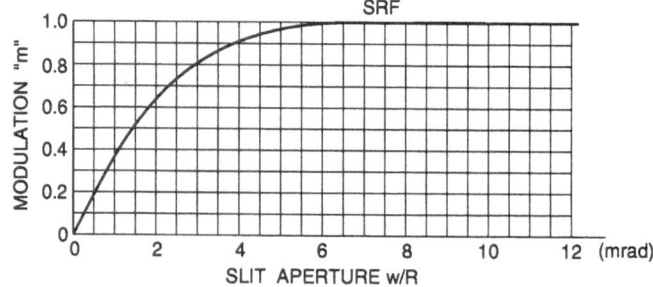

Fig. 3.3. Simple procedure to measure the slit response function (SRF).

For infrared systems, it is important to know the minimum spatial resolution which yields to a signal strong enough to obtain a repeatable measurement and an acceptable detection rate. In this respect, the SRF (slit response function) can be measured. Holmsten (1986) proposed a simple method to derive the SRF (Fig. 3.3). A slit of variable width w is placed at a distance R from the detector; a hotter than ambient background is present behind the slit. The shape of the signal along a line is recorded. A very wide slit will give a profile with a plateau. The amplitude A of the plateau corresponds to a modulation m of 100%. This is the maximum contrast case from all white (hot) to all black (cold), when the infrared image is observed on a black-and-white monitor. The slit width w is then reduced and the new peak height normalized with A gives the corresponding m factor. This technique allows the SRF curve to be plotted (Fig. 3.4).

At this point, a simple example is of interest. Suppose the infrared camera having the SRF curve of Fig. 3.4 has a 12° aperture lens with 105 pixels per line and is placed 1000 mm from a screen. In these conditions the field of view is 210×210 mm (Fig. 3.5).

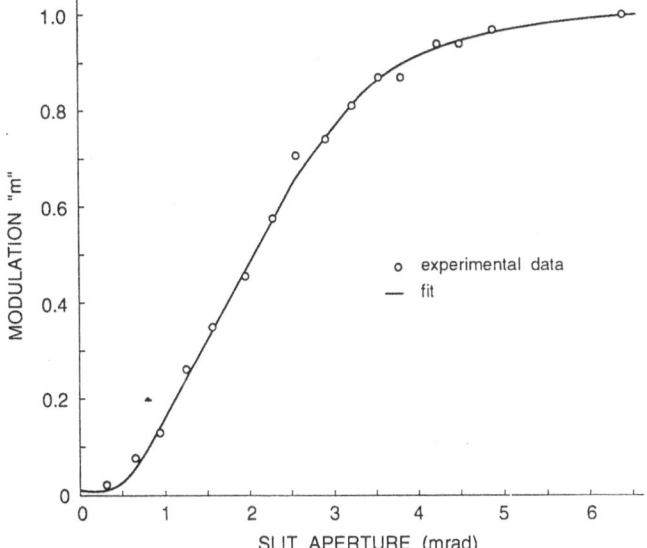

Fig. 3.4. Experimental SRF curve for an AGEMA 782SW.

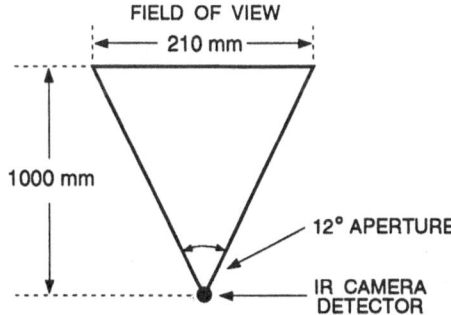

Fig. 3.5. Example of use of the SRF curve to evaluate the spatial resolution of the infrared camera.

The SRF of Fig. 3.4 gives a spatial resolution of 6 mrad at $m = 100\%$ and 2 mrad at $m = 50\%$. If we select $m = 100\%$ that is if we stipulate that a perfect contrast from all white (hot, 255) to all black (cold, 0) is necessary to distinguish between two objects in the image, these objects have to be separated by 1000 tan (6 mrad) = 6 mm. However, if we select a value of $m = 50\%$ (this is generally the accepted value), this means we consider that a 50% contrast suffices to distinguish both objects and they have to be separated by 1000 tan (2 mrad) = 2 mm, this is the *spatial resolution* at $m = 50\%$. These values can be related to the *apparent resolution* which, for our example, equals 210 mm/105 pixels = 2 mm/pixel. In this particular example the apparent resolution corresponds to the spatial resolution, though obviously this is not always the case.

Of course, the more pixels are available (for instance in a bidimensional array), the more likely the spatial resolution will be of acceptable value. We may also note that the apparent resolution may be different along rows and columns if the number of elements in the array is different along both axes. This situation occurs for many infrared radiometers (e.g. AGEMA 880 SW, 175 elements per line and 280 lines per frame at $m = 0.5$). In this case, corrective magnification coefficients have to be applied by software in order to correct the distortion effect in the image (at least for viewing comfort). Finally, it is also of interest to note that the SRF curve can be used to restore the approximate size of narrow observed objects (Beaudoin and Bissieux 1992).

3.1.2 Selection of an Operating Wavelength Band

Selection of an Atmospheric Band

When selecting an infrared camera suitable for TNDE, one major question arises: "Which atmospheric band is to be selected?" Since the atmosphere does not have perfectly flat transmission properties (Fig. 1.9, Chap. 1), the selection of the operating wavelength band will be conditioned by the final application. For the majority of NDE applications, the useful portion of the infrared spectrum lies in the range 0.8 to 20 μm. Beyond 20 μm, applications are more exotic such as high performance Fourier transform spectrometers which operate in the 25 μm range (Koehler 1991). The choice of an operating wavelength band dictates the selection of the detector type as Fig. 1.9 shows. Among the important criteria for band selection are operating distance, indoor–outdoor operation, temperature and emissivity of the bodies of interest.

Since they match the atmospheric transmission windows, the most useful bands are 3 to 5 μm, often called *short* wavelength (SW) band, and 8 to 12 μm, often called *long* wavelength (LW). Most of the infrared commercial products operate in these bands while near infrared (0.8 to 1.1 μm) is easily covered by standard ambient operation temperature silicon detectors (such as standard CCD cameras).

As Planck's law stipulates (Eq. (2.2) and Fig. 2.2), high temperature bodies emit more in the short wavelengths; consequently long wavelengths will be of more interest to observe near room temperature objects. Emitted radiation from ordinary objects at ambient temperaure (300 K) peaks in this long wavelength range. Long wavelengths are also preferred for outdoor operation where involved temperature differentials are small and distances tend to be large. Notice that in the LW band signals are less affected by radiation from the Sun (the Sun emits *less* at the long wavelengths, thus explaining why when using an LW camera, distinction between day and night is difficult (Turck 1988)).

It is interesting to point out the importance of the observation distance. At 100 m, the atmospheric absorption is 0.93 for LW and 0.73 for SW. However, for operating distances restricted to a few metres in the absence of fog or water droplets, the atmosphere absorption has little effect in either band.

Effect of the Emissivity

Spectral emissivity is also of great importance in the choice of an operating wavelength band since it conditions the emitted radiation (Eq. (2.3)). Polished metals with emissivity smaller than 0.2 cannot be observed directly since they reflect more than they emit. A high emissivity coating (such as black paint) or a reflective cavity must be used (Chen et al. 1990; Krapez et al. 1990; Maldague et al. 1991a; Cielo 1992). For metals, emissivity generally decreases when the wavelength of observation increases, thus the SW band will be preferable (Pajani 1987a, b), especially at high temperature (see Planck's law curves in Fig. 2.2).

Effect of the Detectivity of the Detector

Another important point to consider is the detectivity D^* of the detector used (Sect. 3.1.1). As seen in Fig. 1.9, for instance, a 77 K cooled InSb detector operating in the range 3–5 μm has a seven-fold higher detectivity than a 77 K cooled HgCdTe detector operating in the range 8–12 μm. This means that even if, for a specific application, the emitted radiation (temperature of interest, spectral emissivity) is higher in the range 8–12 μm, the contrast obtained may be stronger in the range 3–5 μm because of the superior D^* of the InSb detector (Levinstein 1965).

As a final note, although no specific rule can be formulated, we may point out that detailed studies (Gaussorgues 1984b, pp 421–436; Pajani 1987a,b; Woolaway, 1991) have concluded that for temperatures in the range -10 to $+130\,°C$ measurements can be done without much difference in both bands (3–5 μm and 8–12 μm). For some special applications (e.g. for the military), bi-spectral cameras operating simultaneously in both bands have been developed in order to characterize more accurately thermal signatures of targets.

3.1.3 Acquisition and Analysis Equipment

Figure 3.1 at the beginning of this chapter shows the general set-up of the thermographic NDE inspection station. It comprises typically the following components (Fig. 3.1):

1. Sample holder:
    Part holder
    Moving slide
    Robotic arm
    x–y–z translation stage (or x–y wide table, Dubrovskii et al. 1990)
2. Acquisition head (Fig. 3.6):
    Infrared camera
    Appropriate objective lens (including zoom, close-up, telescope, expansion rings

Fig. 3.6. Acquisition head for a typical TNDE station showing the infrared camera (camera electronics on a trolley, at left), a specimen mounted on the sample holder and the heat stimulation device (two 1000 W projectors with back reflectors).

    which may be needed depending on the application)
    Cooling unit, liquid nitrogen supply, etc.
    Thermal stimulation device
    Tripod, cart
    Battery-operated power supply
3.   Associated electronics and analysis equipment:
    Computer and associated software
    Cart for moving the equipment around
    35 mm or Polaroid™ camera, printer, video cassette recorder (VCR) or digital
    recording system (magnetic or video)
    Battery-operated power supply (12 V car/van, solar power)

Sample Holder

Depending of the size of the parts to be inspected, a sample holder will be necessary. If the objects are large, they may hold in place by themselves or they can be mounted on a moving slide (e.g. Unislide from Velmex, East Bloomfield, NY) for fast inspection of large flat surfaces (Fig. 3.7). In the case of smaller parts, a special holder may be necessary (see Fig. 3.6 for instance).

    A potential difficulty may arise when inspection must proceed horizontally and if the infrared camera is of the liquid nitrogen cooling type since the internal camera dewar (Fig. 3.2) must be held vertically (to avoid spilling of liquid coolant). In this case, a highly reflective mirror set at 45° in front of the camera allows horizontal viewing (Fig. 3.8). First surface gold or aluminium mirrors are almost 100% reflective

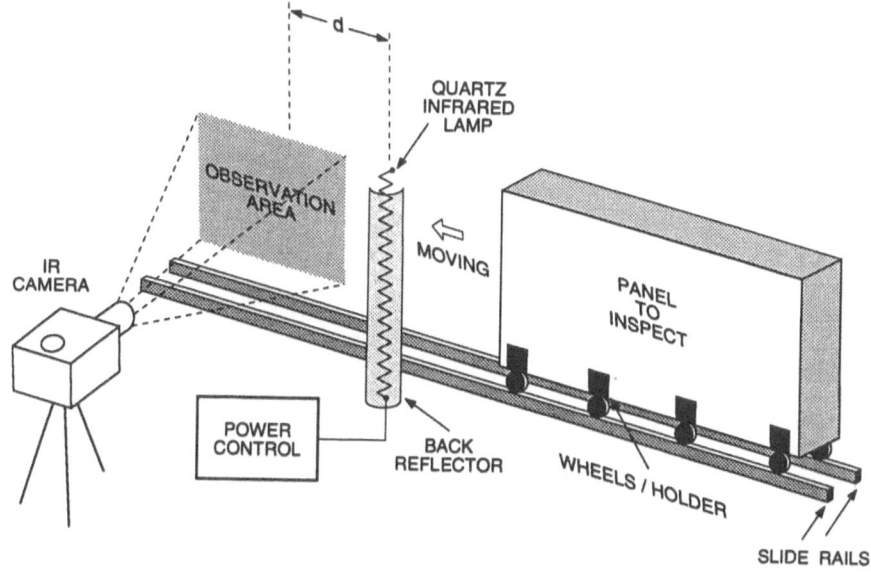

Fig. 3.7. TNDE configuration for the inspection of large components mounted on a moving slide.

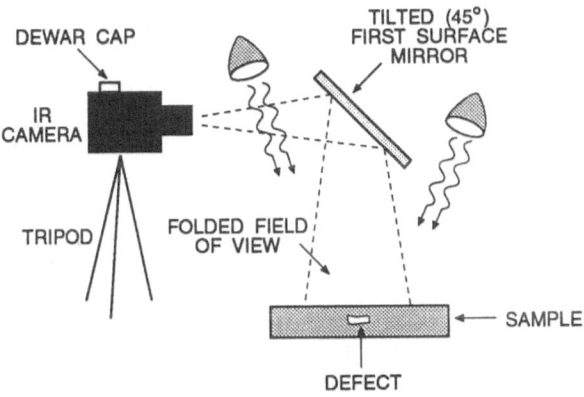

Fig. 3.8. TNDE configuration making use of a tilted mirror for down view inspection.

for wavelengths from 3 μm and up and are well suited to these applications. When using such mirrors, it is important to recall that the recorded images will be inverted.

Acquisition Head

The acquisition head is composed basically of two elements, the infrared camera and the thermal stimulation device. The acquisition head can be either static (e.g. mounted on a tripod) in front of the part to be inspected or mobile, mounted on a robotic arm in the case of inspection of large complex shape workpieces (see example of Sect. 3.1.6 below).

Infrared cameras (radiometers) and detectors were reviewed in Sects 1.4, 3.1.1 and 3.1.2. An example of field-of-view computation is given in Sect. 3.1.6 below. Concerning infrared cameras, we will only mention that unless they are thermoelectrically cooled, special precautions have to be taken to ensure proper cooling (Sect. 1.4), if needed (e.g. a sufficient supply of liquid nitrogen to allow periodic refilling). To give an idea, a well insulated 10 l dewar tank typically keeps up to 10% of its original content over a one-month period (without liquid withdrawal).

Thermal Stimulation Devices, Hot and Cold. Often, lack of success in TNDE is related to problems with the thermal stimulation device such as for instance, the lack of homogeneity of the heating source (Vavilov 1980). Since this question is of great importance, we now concentrate on this subject. The thermal transient needed for part stimulation can be either hot (hotter than the part) or cold (cooler than the part) since a thermal front will propagate in the same fashion, being either cold or hot. The cold approach has, in certain circumstances, several advantages over the warm approach, for instance if the part to be inspected is already hot, such as at the output of a curing oven, in the case of graphite epoxy panels. In these cases, it is uneconomic to add extra heat to the part; cooling is then more economical. Another benefit is that a cool thermal stimulation device will not generate any thermal noise (which may reflect on the part and corrupt the images) since it is at a reduced temperature with respect to the component to be inspected. Such a cool thermal perturbation device can be deployed using cool water jets, cool air/gas jets, snow, ice (with contact), or a cool water bag (with contact). Examples of the cool approach will be given in Sect. 4.3.

More traditionally, hot thermal perturbation is used; this can be high power cinematographic lamps, a bank of quartz line infrared lamps, a high power photographic flash, a laser beam, a heat gun, hot water jets, hot air jets, or a hot water bag (Milne and Carter 1988).

The main desirable characteristics of the thermal perturbation (either cool or warm) are repeatability, uniformity, timeliness and duration. Repeatability is needed in order to compare the results obtained on many identical parts (lamp ageing, for example, will affect this – see below). Stimulation uniformity is necessary when inspection is performed surface wide. Uneven thermal stimulation will produce spurious hot (or cold) spots which may be interpreted as subsurface flaws. Stimulation uniformity is hard to achieve, however. Special deconvolution algorithms (Bumbaca and Smith 1988; Winfree and Welch 1989) or special calibration can allow for stimulation correction after the fact (Maldague et al. 1991c).

Laser scanning of the inspected surface is an attractive possibility although it may complicate the set-up design. It provides, however, high and uniform energy deposit over the surface. A CO_2 laser emitting at 1.06 μm is often used as a stimulation device since the heating beam is not seen by the infrared camera either at 3–5 μm or 8–12 μm (Welch et al. 1990). In this discussion we refer more to imaging although point-by-point inspection is also possible, to determine fibre orientation in composites for instance (Krapez et al. 1987), or to evaluate material physical properties such as thermal diffusivity for instance (Parker et al. 1961). This last application will be discussed in Chap. 8.

Timeliness and duration are other important issues in thermal stimulation. Although step heating is possible, pulse stimulation is generally preferred since it allows one to test simultaneously all the frequencies (flat spectrum of a Dirac pulse) with a smaller risk of overheating the sample. For better agreement with theoretical models and to enable data inversion (Chap. 6), a square thermal pulse is needed. A sliding or rotating shutter may be used in front of the thermal stimulation device to achieve such pulse shaping (Balageas et al. 1988; Delpech and Balageas 1991).

Depending on the part's thermal characteristics, mainly the thermal conductivity K, different pulse durations have to be selected, small (in the few millisecond range) for high conductivity samples and long (a second and over) otherwise (see Eq. (1.1)). Photographic flash (xenon) lamps or a mechanical shutter will produce short high power pulses, which are desirable for inspection of high conductivity parts such as aluminium components (Delpech and Balageas 1991). If requirements of time constraints and heating uniformity are not essential, such as for qualitative TNDE or in the case of inspection of low thermal conductivity specimens, a heat gun or cinematographic lamps can be used for thermal stimulation (as in the experimental set-up of Fig. 3.6).

Line heating is also attractive. In this configuration large panels can be inspected as discussed previously. This configuration may also be reversed; in this way the line heater is mounted on the slide and the part is stationary. The inspection can proceed in transmission (IR camera on one side and heater on the other) or in reflection (IR camera and heater on the same side). Such lateral motion is advantageous since heating uniformity is obtained along the moving direction which may not be the case in a static configuration. Also, such lateral motion allows interesting image processing possibilities by integrating multiple images for noise reduction. This subject is discussed in Sect. 4.3.3.

Contact thermal stimulation such as with ice, snow, or a cold/hot water bag is possible, although it does not allow easy automation or repeatability. However, authors have used successfully lateral propagation surface heating to detect the presence of surface cracks in turbine blades, for instance (Burger and Babak 1985; Vavilov 1990).

When the part is hollow, it may be of interest to use internal stimulation with a flow of liquid (water) or gas (air). In this configuration the change in flow temperature (hot to cold or the reverse) permits discovery of abnormal variations of wall thickness or blocking passages because of the delayed arrival of the thermal perturbation. Following Eq. (1.1), a twofold variation in wall thickness will produce a fourfold variation in the time of arrival of the thermal perturbation on the outer surface after the internal perturbation is initiated. For instance, authors have used this method for jet turbine blade inspection and for wall thickness evaluation in corroded pipes (Beynon 1982; Ding 1985; Bantel et al. 1986; Maldague et al. 1989b, 1990a, c). One advantage of this method is that, since the thermal perturbation circulates inside the part, it does not generate any spurious thermal reflection which may contaminate the measurement as in the case of external heating sources. This method will be reviewed in Chap. 5.

We will now introduce more practical considerations. In the mobile configuration (Fig. 3.9a) line heating can be deployed using a quartz infrared lamp (e.g. the GE QH 1600T, 30 cm long, 1600 W lamp) mounted with back aluminium reflector. Line cooling using a column of air/water jets is an attractive alternative for thermal stimulation, if possible, as mentioned before. However, the contact of water with the workpiece may not be acceptable in some instances.

For cases where the sample is not moving (Fig. 3.9b), a rosette of say six unfocused projectors mounted with a parabolic back reflector around the infrared camera and

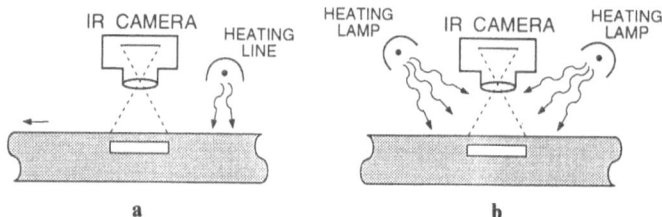

Fig. 3.9. Configurations for thermal inspection: **a** mobile; **b** static.

connected to a timer allows thermal pulses of variable lengths to be obtained (minimum ~ 0.5 s). Careful orientation of the individual projectors within the rosette helps to maximize the uniformity of the heat deposit on the inspected surface (Buchanan et al. 1990). Experimental trials have shown that for a six 1000 W projector rosette the maximum inspected area is about 20 × 20 cm.

Figure 3.10 shows the typical shape of a heating pulse obtained with such incandescent projectors (as pictured in Fig. 3.6); it is characterized by a fast rise time (power on) and slow filament cooling (power off). One problem of incandescent sources used as heat stimulation devices is the warm-up of the mechanical structure which occurs as inspection tasks are repeated. This warm-up causes the lamp structure to emit parasitic thermal radiation reflected on the sample surface, picked up by the infrared camera and superposing with the useful signal, thus causing measurement perturbation (Fig. 2.1). Fans or cooled water circulation help to reduce this effect. Finally, as

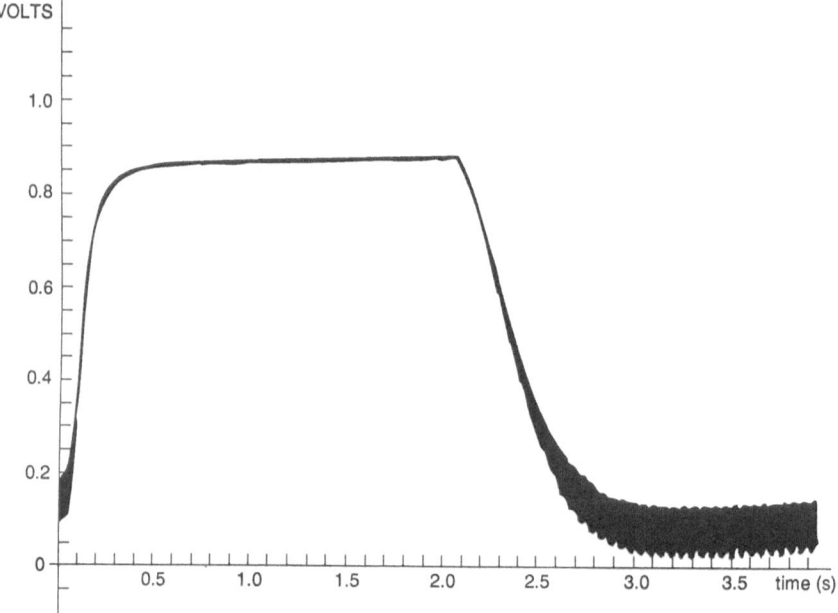

Fig. 3.10. Shape of the heating pulse obtained with a 1000 W incandescent projector (the signal is picked up with a light transducer model A from PhotoAmp, Burbank, CA connected to a data acquisition system DATA 6000 from Data Precision (Analogic Corp), Danvers, MA).

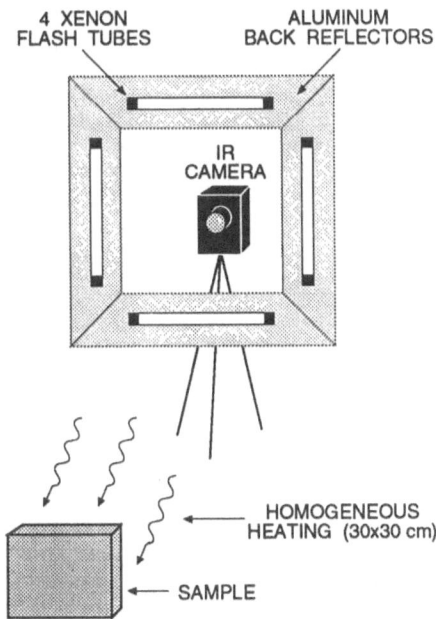

Fig. 3.11. Static configuration; homogeneous heat deposition is obtained by using four xenon tubes with back reflector mounted in a square with camera in the centre. Short heat pulses of less than 10 ms are possible.

noted above, lamp ageing is another drawback which precludes measurement reproducibility (up to 40% of the light intensity can be lost through deposition of filament-evaporated tungsten on the side of the glass envelope).

In order to improve the uniformity of heat deposition in the static configuration of Fig. 3.9b an attractive possibility is to mount four line-heating xenon flash tubes as a square, with the infrared camera placed at the centre (Fig. 3.11): pulse lengths of less than 10 ms are possible with a homogeneous energy deposit of up to 20 kJ over a 30×30 cm area (such a configuration is commercially available from Mecica, 51100 Reims, France).

Another heating technique employs a heating radiator such as the one seen in Fig. 3.12 (Watlow, St Louis, MI, USA, model 8745C, 2500 W, heating surface: 30 cm (h) $\times$ 20 cm (w)) which is mounted on a moving slide in such manner that it heat-stimulates the workpiece while passing in front of it. This kind of stimulation yields a better uniformity of the heat stimulation than with the rosette discussed previously. As an example, given the unit of Fig. 3.12, a deposit of $20 \, \text{J cm}^{-2}$ is obtained at a speed of $4 \, \text{cm s}^{-1}$, $8 \, \text{J cm}^{-2}$ at $10 \, \text{cm s}^{-1}$ and $1.6 \, \text{J cm}^{-2}$ at $50 \, \text{cm s}^{-1}$. Even if this configuration is not really static as for the rosette, it is an interesting approximation for experimental studies (McLaughlin 1985).

Associated Electronics and Analysis Equipment

In addition to the camera electronics, some additional equipment may be needed such as a video timer (to superimpose acquisition time on images), and special

Fig. 3.12. Radiant heater mounted on the moving slide. An aluminium honeycomb specimen is positioned on the sample holder (on left).

equipment to trigger the thermal stimulation devices. It is not really possible to describe such equipment in great detail since it is very much installation dependent.

Concerning the video timer (e.g. some models are manufactured by FOR-A, Japan), it may be convenient to generate timing information on images in a machine readable fashion. This is essential if sequences of images (e.g. temperature evolution over the inspected part) have to be digitized and processed. One way to achieve this is through the use of a bar code superimposed on infrared images. Some IR camera manufacturers (such as Inframetrics) already use this scheme to real-time code pertinent information on the top lines of thermograms. This procedure is convenient since it is only necessary to digitize the infrared image to obtain all its related information (the image itself, thermal range and level, time and date of acquisition, atmospheric coefficient factor, etc.).

A video timer is controlled through a three-input connector for reset, start and stop functions. In the case of a sample moving on a slide, opto-detectors such as TIL-138 (from Texas Instruments) can be actuated by a blade mounted on the sample holder which breaks the emitter–detector beam. Figure 3.13 shows a typical drive circuit for such an opto-detector. The start signal obtained with the circuit of Fig. 3.13 can also be used to trigger the image acquisition sequence.

A computer is needed for infrared image processing. If processing can be done off-line, an Apple Macintosh or a PC-type computer such as an AT, a 386 (with co-processor) or a 486 can be used, although a workstation (such as a Sun Sparc) provides greater processing power and a richer working environment. Much commercial image processing software can be purchased from infrared camera vendors. To control infrared acquisition equipment, a multi-task workstation is not generally recommended since it cannot send and receive signals in real time or with known and repetitive delays. Notice however, such workstations can be operated with real-time operating systems such as with VxWorks™.

If not too complicated image algorithms are needed to assist interpretation by the operator, a portable computer can be used, for example if only display and profile

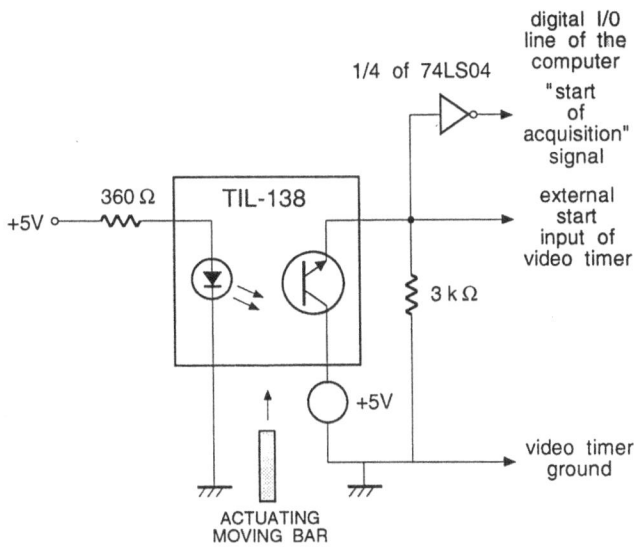

Fig. 3.13. Schematic diagram of the interface circuit needed to interface a TIL-138 opto-coupler to the video timer.

operations are required. In this latter case, an integrated frame grabber–frame processor board plugged into the computer itself will allow greater throughput. Companies such as Data Translation (Marlboro, MA, USA), Dipix (Ottawa, Canada), Matrox and Occulus (Montréal, Canada) or Sharp (Irvine, CA, USA) offer such products.

When very complex computation must be performed over long time sequences of infrared images, especially if close to real-time decisions must be taken (e.g. if inspection proceeds along the production line), then a more powerful processor architecture must be used. This may be the case for quantitative inversion procedures (Chap. 6) or for thermal tomography (Chap. 9). Massively parallel processing machines such as SIMD machines (same instructions – multiple data), are attractive in this context (Krummar 1991). In fact, image processing leads very naturally to parallel processing since pixels can be associated with individual processing elements (PEs). Massively parallel machines such as the recent Maspar MP-1 or MP-2 (Framingham, MA, USA) offer PEs in increments of 1024 but have not been very much used for TNDE applications yet, although this should change in the future.

Battery-operated power supplies may be needed to power sensitive electronic equipment, especially in electromagnetically contaminated environments such as in areas close to high power machinery. In such cases RF shield casings may be required as well. An advantage of DC-operated power supplies is that power fluctuations (such as the ripples in AC–DC conversion) are eliminated. For instance photoconductive detectors require an external current to measure conductivity change. Since a biasing current is needed for photoconductor-type detectors, high charge-capacity alkaline batteries can be used to minimize noise and ripple, which is essential to achieve stable results.

3.1.4 Signal Recording

Video

Although an analogue video signal is often available at the camera output (at least to connect to a TV monitor), some manufacturers propose also a direct digital output which is convenient when further image processing is to be done (in order to avoid multiple A–D, D–A conversions which degrade signal quality because of augmented digitization noise). However, since image format is generally close to the 512×512 picture elements (or pixels), it is generally not possible to record the signal over a longer period of time. For instance, at 30 images per second and 512×512 pixels per image, registration of a 1 min time period requires 450 Mbytes of memory space, which is still considerable. Two solutions are of interest: direct digital registration which can proceed only at a specific time (for instance, using a logarithmic time scale, Sect. 4.3.2) or analogue recording of the video signal on a VCR and later digitized when needed. This latter possibility is attractive since commercial VCRs and video cassettes are available at affordable prices.

If a commercial VCR is to be used for image recording, the user must be aware of two important limitations. First of all, household VCRs are generally built with an automatic gain control (AGC) which cannot be turned off. The effect of this AGC is to control average intensity level of every image so that very bright or very dark scenes can be viewed without saturation. This effect is very desirable for consumers since it adds to the viewing comfort. However, it is detrimental in quantitative thermography, since it rejects the DC level of infrared images, eliminating the possibility to convert image intensity values into absolute temperaures unless special measures are taken, as discussed below.

First of all, in order to correct for the AGC, a grey scale of known values can be superimposed on the images before recording on the VCR. After image digitizing, pixel intensities can be corrected to their original values, based on the comparison between both the recorded and known original values of the grey scale. Of course the process is slow since such processing must be performed for every image (the AGC may change from one image to the next).

As an example of the AGC effect, the values in Table 3.1 have been obtained (the intensity of a white reference square always present in the image is given (Fig. 3.14)).

Although magnetic or magneto-optic recording is actually the only *affordable* means of registering (or archiving) a real-time succession of thermal images, it has, however, some other limitations. The main sources of noise in a magnetic VCR recording unit are found especially at the level of recording and playback electronic circuitry rather than at the tape level unless tape roughness hits the recording head causing bursts of erroneous data (Waggener 1987).

Other aspects of magnetic recording on commercial VCR which must be addressed are the dynamic range and the signal bandwidth. Typical VCR bandwidths are

Table 3.1.

	All white image[b]	All black image[b]
Direct transfer[a]	181	180
Transfer via the VCR (AGC present)	167	145

[a]Digitizing of the image without prior recording on the VCR.
[b]Infrared image area, (Fig. 3.14).

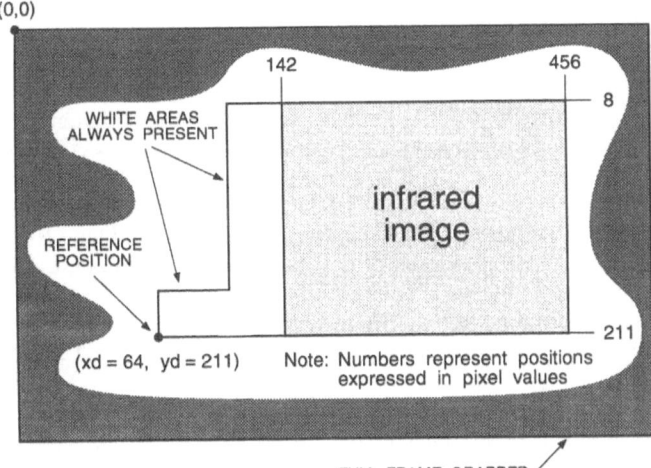

(0,0)

142   456

8

WHITE AREAS
ALWAYS PRESENT

infrared
image

REFERENCE
POSITION

211

(xd = 64, yd = 211)   Note: Numbers represent positions
expressed in pixel values

FULL FRAME GRABBED

Fig. 3.14. Image format for the demonstration infrared station used in this book. The reference point used for image extraction from the VCR is also shown. Note that the raw digitizing format is 512 × 480, although only a subset is used for the image.

around 3.5 MHz, which is enough to accommodate a TV signal of about 300 lines in regular monochrome VHS format (more than 400 lines in S-VHS format). In fact due to the limitations of the electro-optical scanning systems discussed previously, infrared radiometers have generally about 250 active lines (e.g. 200 on the Inframetrics 600 model, 280 on the AGEMA 880 SWB model). Consequently, VCR bandwidth is not a dramatic problem. What is of more concern is the dynamic range. Intuitively, dynamic range can be viewed as the ratio between the smallest and the largest observable signal expressed in decibels (dB). For instance if signals of 5 V and 0.5 mV are to be correctly handled by a particular device, the dynamic range of this device should be at least $20 \log_{10} (5/0.5 m) = 80$ dB. In the case of IR signals, the resolution is generally 8 bits; the required dynamic range is thus $20 \log_{10} (256/1) = 48$ dB (more recent infrared systems operate on 12 bits; they thus require a dynamic range of 72 dB).

If VCR specifications are not available, the following simple experiment can be performed to get an estimate of the dynamic range figure. An image consisting of vertical bars is generated (for instance at the output of a frame grabber, see below). The bars are alternatively of values of 127 and 128 respectively (on an 8 bit frame grabber). The bar image is recorded, and played back on the VCR while the VCR output is observed on a good quality monitor. If the bar pattern is restored without much distortion, we consider the dynamic range is about 48 dB (since the intensity difference in the bar pattern is $(128/256) - (127/256) = 1/256$. Generally, commercial household VCRs have a dynamic range of about 40 dB which thus slightly corrupts (through attenuation of high frequencies) 8 bit infrared images and are unacceptable for 12 bit systems.

Finally, another problem which arises if a VCR is used to record infrared image sequences is the extraction of the useful signal. Because of the limited spatial resolution of the currently available infrared radiometers, the infrared data encoded in the video signal occupies only part of it (e.g. as said previously, only about 250 infrared active lines are available in current systems out of 300 lines available in a mono-chrome VHS signal. This means either infrared lines are duplicated to fill the image or space is left blank or is filled with grey scale and other data. When this video signal is later digitized, it is necessary to extract the useful portion of it (i.e. extract the infrared image) in order to process it, for instance to compute the temperature from the raw thermal data. This problem may be understood with respect to Fig. 3.14. For instance, in this figure, the useful infrared data lies in the area defined by pixels located at corners ($x = 142$, $y = 8$) and (456, 211). Since data extraction is done on the pause mode of the VCR (to digitize a particular frame), the signal will tend to fluctuate. In order to have a correct registration of the extracted infrared data, a software pro-cedure can be used to locate a reference feature in the digitized frame (e.g. the dark-to-white transition point (64, 211) in Fig. 3.14) and knowing the exact distance between this reference point and the useful infrared data, extract and register the image properly.

When extracting images from the VCR, it is possible to improve the signal-to-noise ratio and also restore actual image format by a simple averaging technique. In fact most infrared camera manufacturers "enlarge" the images at the analogue video signal output by duplicating pixels in order to fill the video screen monitor better. This is because of the limited resolution of infrared imagers, as discussed previously. For instance, in the case of the infrared inspection station illustrated in this book, pixels are averaged on a 3 x 3 basis so that the actual digitized image format becomes 68 rows by 105 columns corresponding also to the format of the direct digital image transfer (i.e. not bypassing the analogue video signal). This procedure has also the advantage of saving on the space required to store infrared images in the computer memory and on computational time required to process and interpret these images.

If analogue magnetic recording systems are not suitable (for instance for newer high performance 12 bit infrared systems such as the AGEMA Burst Recording Unit for which a dynamic range of the order of 72 dB is required), magnetic digital disk storage units can be used instead. Such products use Winchester disk drives to create random access to frames and provide in the order of 1 min of real-time recording at 30 frames per second (Kiliski 1991). These systems, although expensive, are advan-tageous since they enable recording/playback without distortion. A digital video disk unit would be another good choice, especially since the video disk cartridge WORM (write once read many) or RWM (read, write many) are removable units which can be exchanged from one inspection task to the next; archiving is then facilitated, and typical capacity is around 650 Mbytes per cartridge.

As a last point concerning video signal recording, we should always be care-ful while propagating video signals through the equipment (VTR, TV monitors, digitizer,...) to avoid signal degradation. Distribution video amplifiers (such as the Panasonic WJ3008) can be inserted in the video path to maintain a correct analogue video signal amplitude and in particular maintain the amplitude of the synchroniza-tion pulses.

Recording Other Infrared Signals

If absolute temperature is to be computed from infrared images, it is necessary to simultaneously record the video signal and also all the infrared camera settings. As mentioned in Sect. 2.1, there are generally two settings of interest on an infrared radiometer: the thermal range and the thermal level (Fig. 2.3). During a given infrared test, the thermal range is generally not changed (to keep the same temperature resolution throughout the experiment) and consequently, only the thermal level is to be recorded along with infrared images. Some manufacturers (such as Inframetrics) combine this information with the video signal. In this case, recording of the video signal suffices to compute the temperature (Sects 2.1 and 3.3). Otherwise, the thermal level needs to be recorded separately.

The thermal level signal can be derived from the infrared camera electronic unit as an analogue signal which can be digitized by dedicated plug-in boards such as the A–D Dash-8 card from Metrabyte (Tauton, MA, USA). When doing so, it is important to match dynamic ranges of both the signal to record and the board. The circuit of Fig. 3.15 allows us to perform such matching. For instance, in the case of our experimental set-up, direct digitizing without using such a conditioning circuit would result in the use of only 84 out of the 4096 possible values available with the Dash-8 board (to cover the thermal level span of interest). The circuit of Fig. 3.15 is formed from three parts: an RC filter to cut off high frequencies ($f_{3dB} = 1/2\pi RC \sim 3.4\,$Hz, high enough to follow the slow rotating action of the operator's fingers on the level knob); gain and offset to bring possible signal variations within the 0–10 V input span of the Dash-8 card: a differential amplifier which allows undesirable ground loop effects to be suppressed.

The thermal level signal is digitized in fixed time increments (such as 50 ms) on a disk file. Since acquisition time is short, it is possible, during one time increment, to average multiple values together in order to reduce the noise level (noise is reduced by a factor of 3 dB every time we double the number of averaged values (Rapaport 1988). It is also possible to smooth this disk file to reduce noise further (cf. Sect. 3.2.2). When temperature processing takes place it is necessary to extract the level value corresponding to acquisition time of the image.

This double recording process (images and level) is necessary on most infrared radiometers. Note, however, that recently, some companies have introduced IR

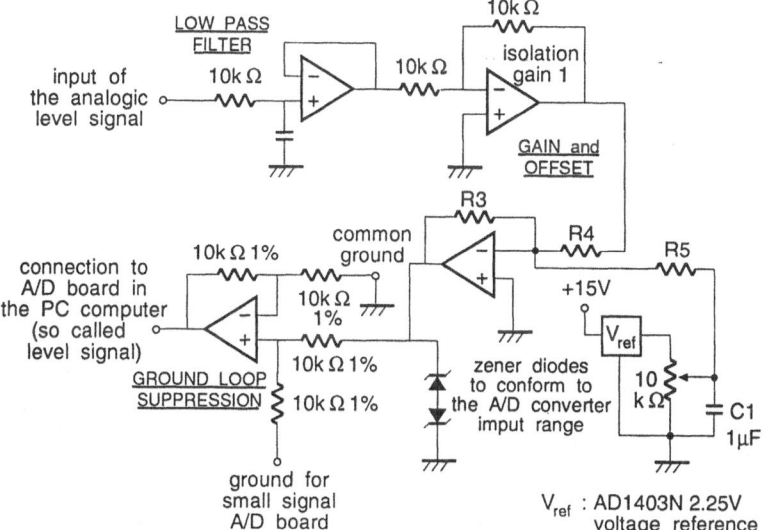

Fig. 3.15. Schematic diagram of the circuit for the conditioning of the level signal; the amplifiers are OP-7 8215 (type 741). The output V_0 is given by (R_3, R_4 and R_5 are 1% resistors): $V_0 = V_{in}(R_3/R_4) - V_{ref}(R_3/R_5)$.

systems with 12 bit digital output (such as AGEMA with the Burst Recording Unit). In this case the dynamic range is so large (72 dB, 16 times greater than more conventional 8 bit systems) that separate recording of thermal range and level is no longer necessary: 12 bit availability allows the whole span of the calibration curve to be covered directly without requiring the *artificial* introduction of level and range (Fig. 2.3).

3.1.5 Measurement Reproducibility

It is interesting to evaluate the question of measurement reproducibility. The question is to determine if we are able to obtain the same absolute values for a given parameter from experiments which take place at different moments. Of course, the experimental conditions, and the workpiece, must be identical. All the instrumentation should also have reached a steady operating regime after a warm-up period of about 3 h. This is particularly important for an infrared camera whose casing acts as a parasitic emitter by itself. This emission has to be taken into account in the calibration process for a complete correction of the radiometric response (Sect. 3.3) and consequently temperature variations of the infrared camera casing will affect the measurement if not yet stabilized. This warm-up period is thus necessary to minimize signal drift (Fig. 3.21).

As an example, Fig. 3.16 shows two typical data sets obtained on two different samples (Plexiglass™ and graphite epoxy plates). For each sample, the same thermographic investigation was repeated seven times over a 2-month period. The measured parameter was the time of occurrence for maximum contrast over an artificial subsurface defect (see Sect. 6.3 for details on contrast computations). For the Plexiglass™ sample the maximum deviation is 4.6% with an average deviation of 1.8%,

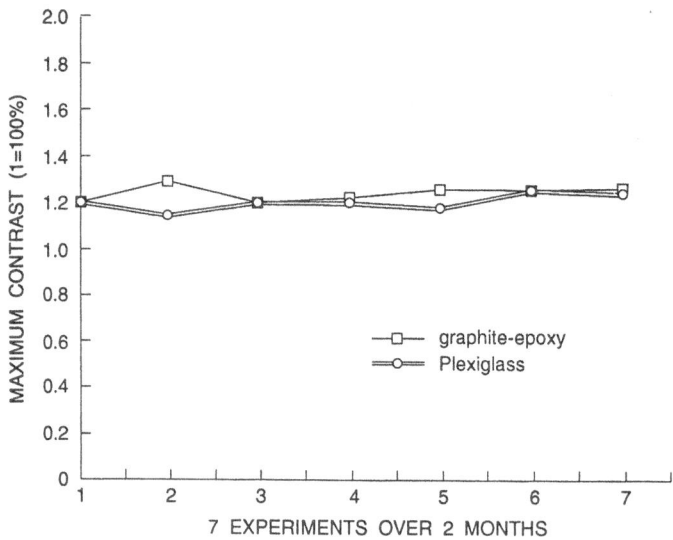

Fig. 3.16. Measurement reproducibility: tests were conducted over a 2-month period. The horizontal scale corresponds to the reproduction of one test. Graph illustrates an important parameter needed for quantitative evaluation, the value of maximum contrast (cf. Chap. 6).

while for the graphite epoxy, we obtain 5.8% and 1.4% respectively. These deviations are in agreement with what is generally expected for IR systems (as discussed in Sect. 2.1).

3.1.6 Example of Analysis for an Experimental Set-up

Before concluding this section, we will study an example which allows us to introduce some interesting considerations about the experimental set-up. When deciding to perform an inspection task by thermography, one of the first concerns is to evaluate the required spatial resolution. A probability detection curve such as the one of Fig. 1.7 may help us to determine the minimum size of detectable defects. The answer to this question allows us to specify the field of view for the inspection task and decide on the experimental configuration.

For example, suppose we want to detect the lack of adhesive in bonded aluminium panels. Suppose the aluminium panels are made of sheets 2 mm thick, epoxy-bonded on a 5 cm foam core (these kinds of panels are used to build thermally insulated cold boxes for trucks). The panels are flat and of large size; the minimum diebonding size to be detected is 25 mm in diameter. Recalling the rule of thumb of Chap. 1 concerning the minimum size of a defect for detection by infrared thermography, that is the defect radius/depth ratio be much greater than one, we see that, for this example, the criterion is satisfied: $12.5/2 = 6.25$, although in this case the refractory properties of the foam strengthen this constraint. With respect to Fig. 1.7, we notice we are over the minimum detection threshold. This brief study indicates that thermography appears a good candidate to detect these defects, providing problems of low aluminium emissivity and high thermal conductivity are solved, for instance using black paint and short thermal perturbations (this kind of sample will be studied further in Sect. 4.3).

We can next evaluate the field of view. In our example, we will make use of the data of Figs 3.4 and 3.5. For a $12°$ lens, without expansion ring, the manufacturer specifies a minimum distance for focus of about 0.7 m. Suppose we install the IR camera at 1 m distance, this gives us a field of view of about $2 \times 1000 \tan 6° \sim 210$ mm. If the image size is 105 pixels wide, a 25 mm wide defect will appear as a blob of about 13 pixels wide (probably less in fact due to tridimensional heat flow effects) or, if expressed as angular width: $25/1000 \sim 25$ mrad. Recalling the SRF curve of Fig. 3.4, it appears such a defect corresponds to a modulation factor of 1 (close to 6 mrad and over is needed for $m = 1$). Consequently contrast detection for this kind of defect should not cause particular problems.

For this example a field of view size of about 21 cm is acceptable (of course larger fields of view can be envisaged). It is, however, important to consider the spatial resolution of the camera as we did here (will the defect contrast be sufficient?); and also will the thermal stimulation device allow a uniform perturbation to be applied over the whole field of view? In the case of inspection of large flat panels, the inspection can proceed on an area-by-area basis. They are many ways to do it. The straightforward method is to divide the panel surface into 21 cm $\times$ 21 cm overlapping areas and then inspect them one by one manually with the thermographic inspection equipment (e.g. infrared camera and associated thermal stimulation device mounted on tripods, associated electronics and analysis equipment carried on a cart). Another possibility is to mount the acquisition head with the thermal stimulation device on a robotic arm which can be programmed to perform the inspection task on a 21 cm $\times$

21 cm area basis (Délouard et al. 1989). This last method is convenient if many identical panels have to be inspected. Another attractive possibility is to maintain the sample on an x–y–z translation stage. This is similar to the robotic arm configuration but with the sample moving instead. The advantage of the robotic arm approach is that larger curved surfaces can be inspected.

Perhaps a faster inspection method is to mount the flat panel on a moving slide with the inspection proceeding in a linear fashion (Fig. 1.4b). Fast inspection rates of up to $1 \, m^2 \, s^{-1}$ can be achieved in this way (Maldague et al. 1987b). This line method can be deployed since defects are always located at the same depth (2 mm, the sheet thickness). This imposes a specific distance between the line heater and the observation area on the surface (see Sect. 1.2.2). It is important to note that this method is impractical if the surface is not flat, in which case other approaches have to be selected (Chap. 10).

3.2 Acquisition Process: Signal Restoration

Infrared system calibration is an important issue (Fraedrich 1991). This section is concerned with the various conditioning operations which are needed to correct image degradation. The only parasitic effects that we will study are those introduced by the instrumentation; other effects such as parasitic reflections, and low emissivity were reviewed in Chap. 2 and will also be examined in Chaps 6 and 7.

3.2.1 Image Degradation

Degradation can occur in three forms: radiometric distortion; geometric distortion; and noise. In a formal manner, we can express the global acquisition process for the position (i, j) in the image by (Arconada et al. 1987; Bumbaca and Smith 1988; Chen 1988; Hershey and Kim 1990).

$$I_{CAM,T}(i, j) = S[h(i, j) * T(i, j)] + n(i, j) \tag{3.2}$$

Where $*$ represents the convolution operator, $S[\cdots]$ concerns the radiometric distortion, $h(\cdots)$ is a linear spatial operator which is concerned with geometric factors and with bandwidth, $n(\cdots)$ is the random noise considered additive, $T(\cdots)$ is the ideal temperature image and $I_{CAM,T}(\cdots)$ is the recorded signal corresponding to the radiance contribution on the calibration curve (cf. Eqs (2.7) and (2.9)). The objective of the restoration process is to recover $T(\cdots)$ starting with $I_{CAM,T}(\cdots)$. This process is complex since nothing, a priori, is known about $S[\cdots]$, $h(\cdots)$ and $n(\cdots)$ which thus must be found experimentally. The correction for radiometric effects seen in Chap. 2 allows us to evaluate $S[\cdots]$ and we obtain

$$I_{CAM,T'}(i, j) \sim h(i, j) * T(i, j) + n(i, j) \tag{3.3}$$

Next, a filter can serve to eliminate the noise while maintaining edges in the image. The resulting image is given by

$$I_{CAM,T''}(i, j) \sim h(i, j) * T(i, j) + n(i, j) - n'(i, j) \tag{3.4}$$

This last equation indicates that a deconvolution is necessary to recover the original image $T(i, j)$ from $I_{CAM,T''}(i, j)$. In the frequency domain, this is expressed by an

Fig. 3.17. Signal restoration steps.

inverse filter. Rewriting Eq. (3.4):

$$\text{IF}''(u, v) \sim \text{HF}(u, v)\text{TF}(u, v) \qquad (3.5)$$

where $\text{IF}''(u, v)$, $\text{HF}(u, v)$ and $\text{TF}(u, v)$ are the Fourier transform of $I_{\text{CAM}, T''}(i, j)$, $h(i, j)$ and $T(i, j)$ respectively. From $\text{HF}(u, v)$, an inverse filter $R(u, v)$ can be found:

$$R(u, v) = \frac{\text{HF}^*(u, v)}{|\text{HF}(u, v)|^2} \qquad (3.6)$$

where $\text{HF}^*(u, v)$ denotes the conjugate of $\text{HF}(u, v)$. This elementary filter considers that $h(i, j)$ is known exactly and that no noise is present; it is also not very efficient near zeros of $\text{HF}(u, v)$ where $R(u, v)$ takes infinity values. We finally obtain

$$\text{TF}(u, v) \sim R(u, v)\, \text{IF}''(u, v) \qquad (3.7)$$

All the restoration steps can be summarized as follows (Figure 3.17):

$$T(i, j) = S^{-1}[I_{\text{CAM}, T}(i, j)] * r(i, j) + n(i, j) - n'(i, j) \qquad (3.8)$$

It is interesting to consider the appropriateness of these operations taking into account their *approximate* character and the needs to be satisfied in the context of TNDE:

1. *Radiometric distortion.* Consider this simple test. A thick brass (high thermal conductivity) plate is brought to an above ambient temperature. It is observed with an infrared camera and an image is recorded. In most cases, the recorded image will not be uniform; moreover, we will observe that the nonuniformity of the image will depend on the plate temperature! Figure 3.24 (see colour section) presents an example of such a test. This simple test indicates that radiometric corrections are fundamental in order to perform quantitative measurements.

2. *Spatial geometry effects.* Consider this simple test. A black painted aluminium plate, on which narrow grooves 1 cm apart are machined on the surface, is observed with an infrared camera and an image is recorded. Figure 3.18 (see colour section) shows a typical image and Fig. 3.19 presents the image profile along a line. From these figures, we notice than the sharp transition of a groove span over 3 pixels instead of 1 (this is related to the apparent resolution discussed

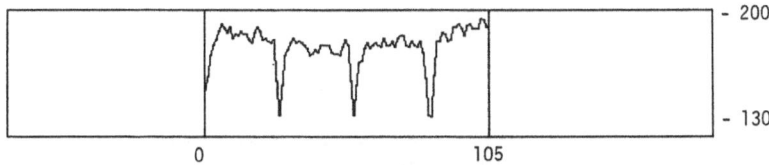

Fig. 3.19. Profile along a row (row 38) of Fig. 3.18; the four grooves are clearly distinguished. Horizontal scale: pixel coordinates along the row, vertical scale: image intensity.

in Sect. 3.1.1). The temperature of the plate is close to the temperature of the camera casing in order to limit the radiometric distortions which are, however, visible in Fig. 3.19. For TNDE applications, important image features (defects) appear as smooth transitions in the temperature images due to the diffusion process as seen in Figs 2.7 and 3.20 (and also in figures presented later in the book). For these applications, spatial geometry effects are not especially significant. In some cases where sharp temperature transitions must be observed, such as for the study of aerodynamically heated structures (e.g. missiles atmosphere re-entry), corrections based on the LSF and SRF notions, have been reported (Wong 1982; Balageas et al. 1988; Beaudoin and Bissieux 1992).

3. *Noise effects.* Consider an infrared defect image recorded on a typical active TNDE scene with an infrared radiometer, a three-dimensional plot is likely to resemble Fig. 3.20. This figure reveals a strong high frequency noise level which must be taken into account.

This study of thermogram degradation allows us to rewrite Eq. (3.8):

$$T(i, j) \sim S^{-1}[I_{CAM,T}(i, j)] + n(i, j) - n'(i, j) \tag{3.9}$$

We will use this relation to restore infrared images prior to their analysis. Radiometric effects, that is spectral response effects from the camera detector, optical path differences, entry pupil, etc. were globally studied in Chap. 2 (Eq. (2.11)); the calibration process required will be reviewed in Sect. 3.3. The noise effect will be the subject of

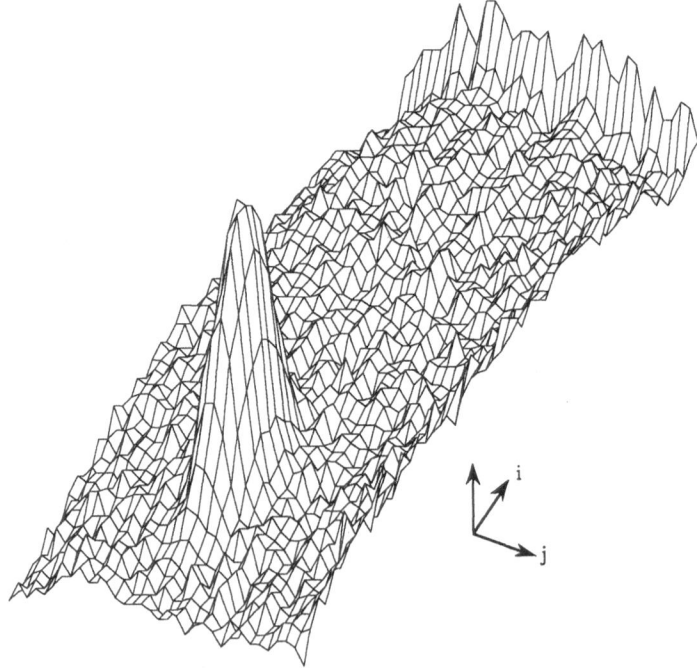

Fig. 3.20. Three-dimensional plot of a raw image recorded over a Plexiglass™ plate (subsurface defect 5 mm diameter, 1.12 mm depth; see Chap. 6 for more details on the quantitative characterization procedure). Note that this graph is drawn with hidden lines using a procedure described by Newman and Sproull (1979, p 372). The higher values on top are due to uneven heating.

the next section. Before concluding this section, we would like to cite Fred E. Nicodemus (1967, p. 289) who advises those who are too confident concerning these "corrections" which can be applied to recorded signals:

> Although often termed "corrections", they are unvoidable uncertainties in these transformations so that the raw measurements are usually more accurate or "correct" (when rightly interpreted as conditions at the instrument) than are the final "corrected results".

3.2.2 Noise

In electronic vision systems, many different types of noise can be found (Ravich 1988) and are depicted in Fig. 3.21: drift, fluctuation, fatigue (e.g. loss of detector sensibility due to repeated radiation exposure), lagging, noise amplification, quantization noise, noise due to electro-optical scanning (for an IR radiometer). To this *system* noise, particularly for IR systems where signals are naturally weak (as discussed in Chap. 1), many adverse effects contribute to further degrade the signal such as: parasitic emission, thermal reflections, random arrival of photons etc.).

At the infrared detector level, we notice essentially (Levinstein 1965): Johnson or thermal noise (present in conductors and due to the random motion of charges in a solid), flicker of $1/f$ noise (present in semiconductors, it depends on the observation frequency), shot noise (caused by the random and discrete photon arrivals in the incident radiation).

For IR radiometers, at the end of the acquisition process, at the pixel level, the noise is additive, of Gaussian nature and of high frequency with respect to the useful signals (Herby 1988). An increase of the image exposure time, thus an averaging of consecutive images, will allow a reduction of these adverse effects. In this case the improvement of quality is a function of the square of the summed images. This

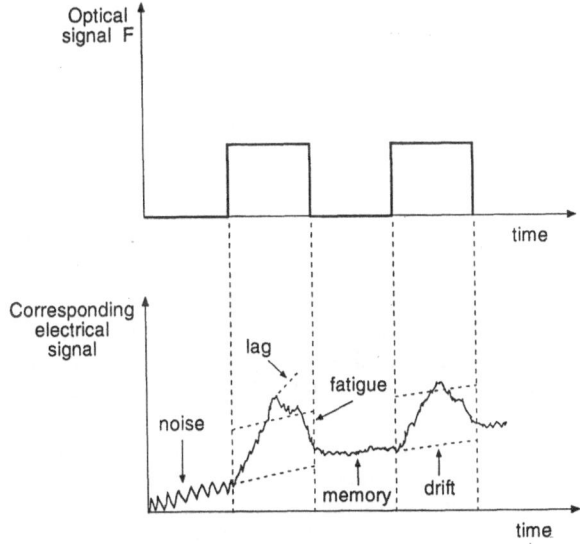

Fig. 3.21. Typical signal degradation processes.

averaging technique can be used whenever it is possible since it is simple and preserves edges, which is not the case when averaging is performed within the image itself (such as with local averaging in 3×3 kernels).

Noise Evaluation

It is of interest to characterize the noise content present in infrared images using a technique proposed by Lee et al. (1987, p 297) and Haddon (1988, p 197). It is only necessary to record two images A and B of the same scene: if they are recorded in the same conditions the only difference between them is noise, which can thus be evaluated. The signal-to-noise ratio (SNR) can be computed using the following formulae:

$$\text{SNR} = \frac{\text{average power image}}{\text{average power noise}} = \frac{\sum_i \sum_j [A(i, j)]^2}{\frac{1}{2}[\text{standard deviation} \langle N(i, j) \rangle]^2 \, \text{Maxcol Maxrow}}$$

(3.10)

where

$$N = \text{noise} \sim |A - B|$$

$$\text{standard deviation } (N) = \sqrt{\frac{\sum_i \sum_j [N(i, j) - \eta]^2}{\text{Maxcol Maxrow}}}$$

η = average of N

$i = 0, 1, \ldots, \text{Maxrow-1}$ (Maxrow rows in images)

$j = 0, 1, \ldots, \text{Maxcol-1}$ (Maxcol columns in images)

The 1/2 factor is introduced in Eq. (3.10) because the grey level variance for the difference image is equal to twice the noise variance of the individual image, considering a Gaussian noise distribution (Haddon 1988, p 197, Eq. 24). As an example of this method, repeated tests gave an average SNR of 32 for our thermal imaging system. As a comparison, the same procedure used with a CCD video camera (Panasonic WV CD-50) gave a value of 1330. As expected, the noise is more important in the IR imager.

   Another method to evaluate noise consists of recording a sequence of images (e.g. ten) over a uniform temperature target whose temperature is as close as possible to the camera casing temperature in order to reduce radiometric effects. The grey level standard deviation is next computed for all the images; moreover, for every pixel location (i, j), the maximum and minimum grey level intensities within the image sequence are found and the max-min difference is computed. This test is next repeated for many different values of the thermal level (Sect. 2.1). Typical results are shown in Table 3.2.

   We notice that noise is independent of thermal level, thus of the intensity; this is of course desirable. Considering this example, an error of $I \pm 5.2$ is associated with a given pixel (i, j) of intensity I. For the mid-scale value, this corresponds to an error of 4% which is in agreement, for the illustrated system, with reproducibility results presented in Sect. 3.1.5.

Table 3.2.

Thermal level	Average intensity	Std dev. intensity	Max-min difference
1657	164	5.68	10.43
1687	162	5.47	10.43
1697	146	5.8	10.23
1735	110	5.79	10.19
1771	78	5.57	10.45
1789	77	5.79	10.34

Noise Processing

To reduce noise effects, when a temporal average is not possible, different image processing techniques can be employed. One of the more common methods consists of dividing the image into small areas such as 3×3 and replacing the central pixel value by the average of the pixel intensities within the kernel. However, this method smooths the edges and thus further distorts the spatial content of the images.

To avoid this problem, images can be processed using well-known median filtering techniques (Narendra 1981; Kapur et al. 1985; Anthony et al. 1987; Nutter et al. 1987; Barnsley and Sloan 1988; Bumbaca and Smith 1988; Lin and Willson 1988). Using this method, the central pixel value of the kernel is replaced by its median value. For instance if the 3×3 kernel has the following values (sorted in ascending order): 64 65 66 67 70 71 72 73 113, the central pixel will be replaced by 70. This method has the advantage of preserving edges while rejecting high frequency noise.

Another efficient method of smoothing images employs a sliding Gaussian window (Hayden 1987). For simplicity, we will first present the unidimensional case.

Let $v(\Omega)$ be the raw signal delivered by the system (vector of values to be smoothed, where Ω is the index variable) and $f(\Omega)$ the transfer function of the system (comprising global effects from the camera, etc.). What is important to notice here, is that the system cannot produce *by itself* frequencies greater than those specified by its function $f(\Omega)$. In this sense, any signal whose frequency is greater with respect to $f(\Omega)$ will be called *inconsistent* with the system and will be have to be eliminated.

The signal $v(\Omega)$ can be expressed as $v(\Omega) = v_n + n(\Omega)$, meaning that, in the nth iteration, the signal is equal to the summation of the ideal (smoothed) signal $v_n(\Omega)$ with the noise $n(\Omega)$. Since the noise is considered statistically uncorrelated with the system function, we have

$$f(\Omega)*n(\Omega) = 0 \quad (* = \text{convolution operator}) \tag{3.11}$$

For the first approximation, we suppose

$$v_1(\Omega) = f(\Omega)*v(\Omega) \tag{3.12}$$

is equal to the ideal smoothed signal. The noise is thus given by the difference between the raw signal and the ideal signal in the first approximation:

$$n(\Omega) = v(\Omega) - v_1(\Omega) \tag{3.13}$$

If we suppose an iterating convergent process towards the ideal (smoothed) signal, we can thus assume that, at the nth approximation, we will have an ideal signal made up of the ideal signal of the previous approximation v_{n-1} plus *residual noise*:

$$v_n(\Omega) = v_{n-1}(\Omega) + \underbrace{[v(\Omega) - v_{n-1}(\Omega)]}_{\text{residual noise}} * f(\Omega) \tag{3.14}$$

Since the number of iteration steps is large and following Eq. (3.11), the residual noise reduces to zero and a larger number of iterations will not degrade the signal, besides eliminating its *incompatible* content within the first few iterations. Practically, two iterations are generally sufficient.

This smoothing process was revealed to be fast and efficient. For images, it is sequentially applied along rows and columns. The $f(\Omega)$ function is given by the normal curve, expressed in continuous form, given by

$$f(\Omega) = \frac{1}{\sigma\sqrt{2\pi}} \exp\left[-\frac{(\Omega - \mu)^2}{2\sigma}\right] \tag{3.15}$$

with μ the average and σ the standard deviation of the distribution and for the discrete case:

$$f(i) = \frac{C}{B} \exp^{-1/2}\left[\frac{i - (i_{max}/2)}{B}\right]^2 \tag{3.16}$$

with

i_{max} = number of elements of the Gaussian
$i_{max} = 10R + 1$, value typically established through testing
$\quad i = 0,1,2,\ldots i_{max}$
$\quad B = 5$ (for a 512×512 image, typically)
$\quad B = 4$ (for a 100×100 image, typically)
$\quad B = 2$ (for a 70×70 image, typically)

The C factor, Eq. (3.16), is adjusted so that the summation of all the Gaussian elements yields 1:

$$C = \frac{B}{\displaystyle\sum_{i=0}^{i_{max}} \exp - \frac{1}{2}\left[\frac{i - (i_{max}/2)}{B}\right]^2} \tag{3.17}$$

This avoids the presence of an undesirable gain: a uniform signal will pass through the process without being affected. The discrete convolution operation $v'(\Omega) = v(\Omega) * f(\Omega)$ of Gaussian $f(\Omega)$ with vector $v(\Omega)$ of N elements gives, for each element Ω of $v(\Omega)$, the new element $v'(\Omega)$:

$$v'(\Omega) = \frac{1}{i_{max} + 1} \sum_{i=0}^{i_{max}} \left[v\left(\Omega - \frac{i_{max}}{2} + 1 + i\right) f(i)\right] \tag{3.18}$$

$$v'(\Omega) = \frac{1}{i_{max} + 1} \sum_{i=0}^{i_{max}} [v(\text{index}) f(i)]$$

where index $= \Omega - (i_{max}/2) + 1 + i$, and for the limit cases:

$$\text{if index} < 0 \rightarrow \text{index} = 0$$
$$\text{if index} > N \rightarrow \text{index} = N$$

Typically, the Gaussian curve is first initialized using Eqs (3.16) and (3.17), then the smoothing is done in two passes using Eqs (3.12), (3.14) and (3.18). The parameter

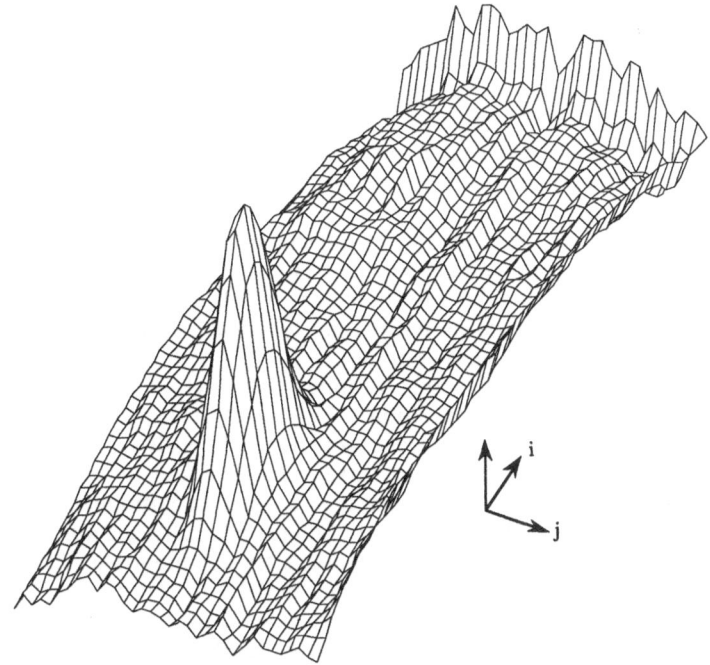

Fig. 3.22. Three-dimensional plot of the smoothed image of Fig. 3.20.

B is defined to obtain an optimal smoothing: this is the classical trade-off between the maximum reduction of the random noise and the minimum deterioration of data (Girard and Algazi 1985). This technique leads to an improved SNR of 152 in the case of the infrared images of the inspection station used as an illustration in this book (instead of 32 for raw data as stated at the beginning of Sect. 3.2.2). This five-fold improvement is due to the high frequency noise content (1 cycle ~ 4 pixels) of our raw infrared images for which the smoothing process is found to be particularly efficient: Fig. 3.22 shows a smoothed image (the raw image is pictured in Fig. 3.20). The interested reader will find the smoothing procedure listing in App. B.

3.3 System Calibration

The discussion of Sect. 3.2.1 insisted on the necessity to process thermograms in order to correct them for radiometric distortions (Eq. (3.9)); that discussion was related to the study of radiometry presented in Sect. 2.1. We will now expose this matter in more detail. More specifically, we will study all the calibration procedures needed in order to recover exact temperature from raw measurements in the case of an infrared radiometer. Recalling Eq. (2.11), we had

$$T(i,j) = S^{-1}\left\{R(i,j)\left(\frac{I_{\text{img}}(i,j) - 128}{256}\right)\text{range} + P(\text{level}) + A(i,j)\right\} \qquad (2.11)$$

which allows us to compute object temperature $T(i, j)$ for every pixel (i, j) in the field of view from thermogram I_{img} and from the thermal level signal. Distinction between I_{CAM} and I_{img} for limited 8 bit dynamic range systems was explained in Sect. 2.1 and can be understood with respect to Eq. (2.9) and Fig. 2.3. In the correction process, it is necessary to evaluate $P(\text{level})$, $A(i, j)$, $R(i, j)$ and $S^{-1}(\cdots)$. In this study, we will illustrate the particular case of our radiometer, as an example. Note that although the actual value of each parameter is specific to a given apparatus, the method to find the parameters is general.

We start with $P(\text{level})$. Plotting the relationship between the level and pixel values allows us to determine the degree of the $P(\text{level})$ polynomial; in our apparatus, this is a first order polynomial (note that this should be the case for all infrared radiometers). Table 3.3 indicates typical values; notice that the slope pixel(level) is independent of the temperature, this is of course desirable.

Values in Table 3.3 were recorded as follows. A black body was set at a given temperature $(10, 25, \ldots °C)$ and positioned in front of the camera. For each black-body temperature, images were recorded for two extreme positions of the "level knob" leading, for the same scene, to different "pixel values" because the maximum observable temperature differential is moved at two different positions on the camera calibration curve (Fig. 2.3). Readings are obtained in a restricted area at the centre of the image where radiometric distortion effects are less important. The selection of a central reference area within the thermogram (10×10 pixels averaged together) ensures a reduced signal degradation for which case we may suppose $A \sim 0$ and $R \sim 1$ (in Eq. (2.11)). Signal degradation is partly due to the differences in optical paths from one pixel to the next and to transmission factors of the optics whose antireflection coatings are optimized for normal incidence.

Moreover, a plot of the black-body temperature as a function of the combined signal corresponding to the argument of $S^{-1}(\cdots)$ in Eq. (2.11) reveals, in the restricted temperature span of interest for thermographic NDE applications (most often between $15 °C$ and $60 °C$), that $S^{-1}(\cdots)$ is a second order polynomial (Fig. 3.23).

In Fig. 3.23 we show the combined signal $[0.8063 * \text{level value} + \text{normalized pixel value}]$ as a function of the black-body temperature. This expression corresponds to term ARG in Eq. (3.19) below when no radiometric distortion is present (i.e. at the image centre). A second order fit is plotted on the same figure and shows the good correspondence with experimental data. In Fig. 3.23, the factor 0.8063 is an estimation for $P(\text{level})$ and corresponds to the average sope for the family of curves "pixel intensity vs level" in Table 3.2.

Table 3.3. Pixel–level relationship

Black-body temperature (°C)	Digitized level signal (0–4095)	Averaged pixel value (0–255)	Corresponding slope of curves pixel (level)
10.0	1195.8	67.3	0.8037
	1036.7	195.1	
25.4	1884.6	65.7	0.8079
	1720.9	198.1	
36.0	2528.6	76.9	0.8105
	2375.1	201.4	
Ambient	1732.5	72.4	0.8035
	1575.3	198.7	

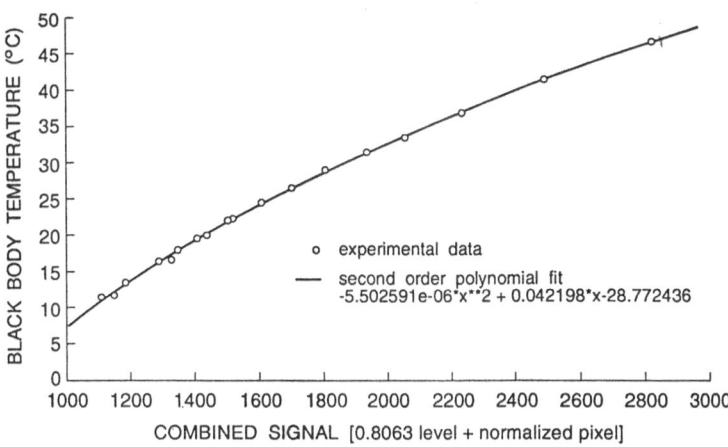

Fig. 3.23. Fitted curve associating black-body temperature to measured camera signal [0.8063 × level value + normalized pixel value] for a second order polynomial. Circles correspond to experimental data. Pixel intensity corresponds to the average value recorded in a small window defined at the centre of the image to limit radiometric distortions (at this location: $A = 0$, $R = 1$).

This study shows that a second order equation for the temperature corresponds well to the camera calibration curve on a restricted span of temperature. The general procedure is as follows. From Eq. (2.11) we can express $S^{-1}(\cdots)$ for a position (i, j) within the iamge by first defining the parameter ARG:

$$\text{ARG} = R(i, j)\left(\frac{I_{\text{img}}(i, j) - 128}{256}\right)\text{range} + l_1\text{level} + l_0 + A(i, j) \qquad (3.19)$$

and then expressing the temperature $T(°C)$ as a second order polynomial in terms of ARG:

$$T = c_2\text{ARG}^2 + c_1\text{ARG} + c_0 \qquad (3.20)$$

The next step is to find the calibration parameters:

$$l_0, l_1, c_0, c_1, c_2 \qquad (3.21)$$

Once these values and matrixes A and R are known, the temperature computations are peformed with Eqs (3.20) and (3.21) with inputs $I_{\text{img}}(i, j)$, the pixel intensity at (i, j) and the level signal for the considered image. The calibration is done in two phases. In the first phase we assume there is no distortion phenomenon at the image centre, so that conditions of Eq. (2.10) are respected, thus simplifying Eq. (3.19).

The actual procedure is as follows. The infrared camera is pointed on the black-body surface whose temperature is increased by steps (typically 5 °C per step) from 15 °C to 60 °C. For each temperature, two cases are recorded for two extreme values of the "level knob". For each case, 15 images are averaged together and the average intensity in a restricted area at the image centre is kept in the computer memory as well as the level value and the true temperaure T_{true}. These values (two per temperature) are passed to an optimization program in order to obtain the parameters of Eq. (3.21) using a mean square minimization of $|T_{\text{computed}} - T_{\text{true}}|^2$, between T_{true} the temperature recorded by the thermocouple connected to the black body and T_{computed} obtained using Eqs (3.20) and (3.21).

In the second phase of the calibration process, we evaluate the radiometric distortions. A thick ($2 \times 25 \times 25 \, \text{cm}^3$) black-painted ($\varepsilon \sim 0.9$) brass plate (the high thermal conductivity of brass ensures a uniform surface temperature) is mounted in front of the infrared camera. The plate is uniformly heated (or cooled) by immersion in a water tank. Three cases are considered:

$k = 1$, plate temperature less than the temperature of the camera casing ($T_{\text{plate}} \approx 15 \, ^\circ\text{C}$)

$k = 2$, plate temperature similar to camera temperature   ($T_{\text{plate}} \approx T_{\text{camera}} \approx 25 \, ^\circ\text{C}$)

$k = 3$, plate temperature higher than camera temperature ($T_{\text{plate}} \approx 45 \, ^\circ\text{C}$)

Figure 3.24 (see colour section) shows three raw images corresponding to these three cases in which the distortions within the image are easily seen. A check can be made for plate temperature uniformity and convection effects by moving the camera field of view across the surface, tilting the plate slightly and observing that the isotherm distribution is constant. It can be seen that when the brass plate temperature is close to the camera temperature, distortions are less apparent because self-emission effects are much reduced. The exact temperature of the plate T_{true} is measured with a thermocouple inserted just below the plate surface.

An average of say 15 images for each case is obtained to minimize the convective loss effects and to ensure that the brass plate is still at uniform temperature. The three

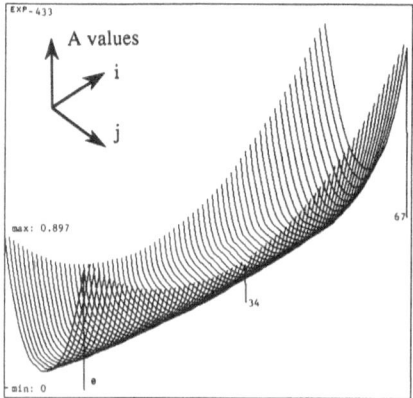

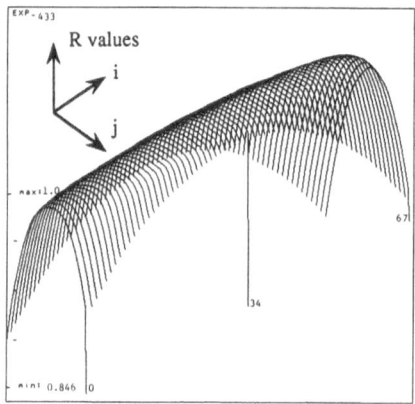

Fig. 3.25. Three-dimensional plot of matrixes A (max 0.897, min 0.0) and R (max 1.0, min 0.846).

images are used by the optimization program in order to compute correction matrixes A and R by minimization of the quadratic error $E_{A,R}$:

$$E_{A,R} = \sum_{k=1,2,3} (T_{\text{computed},k} - T_{\text{true},k})^2 \tag{3.22}$$

by adjusting $A(i,j)$ and $R(i,j)$ values for each position (i,j) considered individually in Eq. (3.19). This procedure allows us to obtain all the parameters for the whole image. Figure 3.25 depicts typical three-dimensional plots of the two correction matrixes; the edge effects are particularly evident. Figure 3.26 shows the effect of the corrections on two images for cold and warm condition of the brass plate. The maximum computed temperature differentials $|T_{\text{computed}} - T_{\text{true}}|$ are shown in Table 3.4.

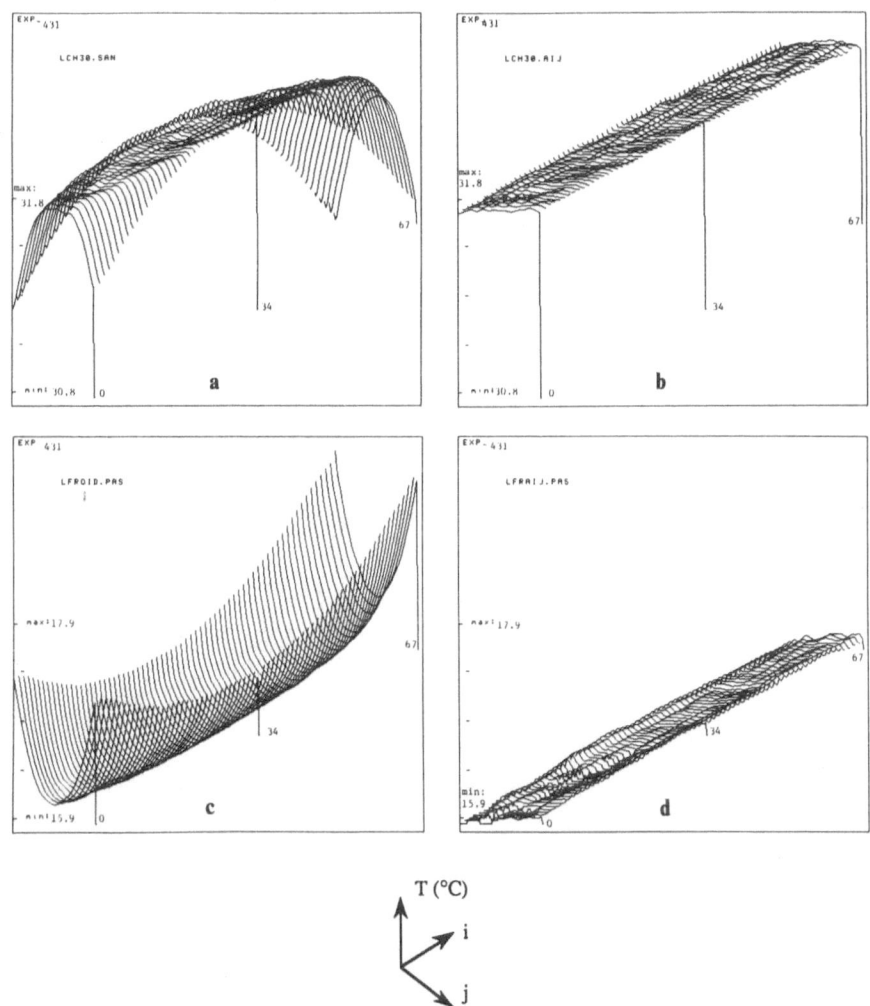

Fig. 3.26. Effects of radiometric corrections, Hot image, **a** before correction; **b** after correction. Cold image, **c** before correction; **d** after correction. Vertical scale is temperature (°C).

Table 3.4.

Brass plate	Before correction (°C)	After correction (°C)	T_{true} (°C)
"Warm" image	1.06	0.18	31.7
"Cold" image	2.04	0.65	16.1

For the "cold" image, a larger differential remains after correction. This may be explained by the fact the "cold" image was recorded some 24 h after the actual calibration took place. Although in that case the correction seems not as accurate, it is able nevertheless to reduce radiometric distortion by at least 50% over the span of temperature (10–50 °C), not considering also the corrections over the whole image.

Before concluding this section, it is important to recall that before any repeatable measurement can be done, the thermographic equipment must warm up for two to three hours in order to reach its steady thermal operating state. This is important since the calibration process takes into account the self-emission of the camera. Also due to ageing, it may be advisable to recalibrate the camera periodically (every 6 months or so). Moreover, the same black paint of high emissivity ($\varepsilon \sim 0.9$) should be used in the subsequent experiments in order to be able to neglect this factor (Eqs (2.6) and (2.7)).

Conclusion. In this chapter, we have studied the thermographic inspection equipment, reviewed its performance and described techniques to restore the temperature signal. In the next chapter we will start to study in more detail experimental techniques and image processing applied to TNDE.

Chapter 4

External Thermal Stimulation: Methods and Image Processing

4.1 General Considerations

As discussed in the previous chapters, the TNDE principle employed to detect subsurface defects consists of submitting the workpiece being inspected to a thermal pulse and analysing the temporal temperature response of the workpiece surface in order to discover eventually anomalies linked to subsurface flaws. As reviewed in Chap. 1, the optimum time of observation is proportional to the square of the depth z_{def} for the defect considered and the TNDE method is especially sensitive to relatively *large* defects (with respect to the defect depth). In this chapter, we will focus our attention to the *classical* experimental techniques and image processing methods needed for defect detection and localization based on this principle. More specifically, we will make use of thermal sources which are *external* to the material, while in the next chapter, we will study applications where thermal sources are *internal* (i.e. located inside the workpiece).

Images have always been an efficient way of representing a large quantity of information; a popular proverb says that "an image is worth a thousand words". In fact the most ancient graphical quantitative representation known is a map of Northern Mesopotamia (now the Anatolia region in Turkey) found on an clay tablet dating from circa 3000 B.C. (Beniger and Robyn 1978). Today and more in the future, electronic images will play an essential role in society as a vehicle for knowledge transmission (e.g. multimedia applications, computer-assisted learning, etc.).

As mentioned in the previous chapters, recorded TNDE images are often corrupted by various sources of noise. Moreover, in the case of the active approach, signals are of small amplitude, and further degraded by temperature spreading due to the three-dimensional diffusion of the thermal front under the surface (as seen for instance in Fig. 2.7). Considering this, special processing is needed in order to enhance TNDE image contrasts, either for the quantitative characterization, for the traditional operator-assisted procedure (e.g. in order to present images in a more comprehensive format) or for automated inspection.

4.1.1 Automatic Versus Manual Inspection

Even after over 30 years of evolution, numerical image processing and automatic inspection are not yet really widespread in the industry (Coster and Chermant 1989; Trivedi 1990). However, this situation is changing now for the following reasons:

Stricter quality requirements dictate quality control implementation at the production stages.

Labour costs are skyrocketing, especially in developed countries. This makes operator-independent automated inspection systems an affordable, if not essential, solution.

Human operators cannot sustain fast production rates (humans are limited to about two or three decisions per second at best and cannot maintain such a rate over long periods of time).

The so-called "penguin phenomenon" by which a company will be forced to follow its competitors who have acquired state of the art technology. State of the art technology can also be an effective marketing tool.

Reduction in the cost of computers, sensors and related hardware.

Reported applications of automatic inspection are still very primitive and often based on a "go/no-go" basis, at least when compared to reported research activities, despite the fact automatic inspection techniques are advantageous. In fact, automatic inspection techniques are advantageous for the following reasons (Freeman 1986; Bieman 1988): precision; uniformity (human interpretation is subjective and prone to inconsistency from one decision to another); no in-depth training of employees required; no performance degradation through boredom and fatigue; relative ease of operation in dangerous or hostile environment conditions; possibility to record the measurements and use them to compile statistics on the production process. In this respect, in Sect. 4.4 we will review some image processing techniques well suited to the application of automatic inspection to TNDE.

4.1.2 False Colour Image Coding

One of the simplest image processing techniques to enhance contrasts is to display images in false colours rather than in grey levels (Fiorini et al. 1982; Bouchardy et al. 1983; Patel et al. 1991). This method is based on the fact the human eye is able to distinguish only a few tens of grey levels, while it can resolve thousands of different colours. This is because, in the human eye, colour-sensitive cone cells are connected to their own nerve on a one-to-one basis, while intensity-sensitive rod cells are connected in groups to optical nerves, thus reducing the amount of detail discernible by these receptors (Gonzalez and Wintz 1987).

This simple method of false colour display is useful but limited and while sometimes it gives good results, it can be very annoying in some instances, particularly when infrared images are taken from visible structures (e.g. a human face, a tree,...) and have to be compared in some way with corresponding visible images. In the case of subsurface defect images, no visible equivalent exists and thus false colour coding is a useful presentation tool which, moreover, allows the noise effect to be reduced, especially if a limited number of colours are displayed (noise is integrated in the various colour bands). In this book, figures illustrating the discussed methods will be presented using this false colour scheme.

4.1.3 Different Types of Materials

In the following sections we have chosen to present the various techniques for investigation, detection and localization based on the type of material inspected. This is an arbitrary choice which is, however, advantageous since it allows us to introduce the techniques with an increasing degree of complexity. First, TNDE will be applied to graphite epoxy workpieces (Sect. 4.2) and then to aluminium bonded laminates (Sect. 4.3). We recall discussion on these new materials was presented in Sect. 1.3. The particular choice of these two materials is based on the fact they have widely opposing thermal behaviours and properties, notably the thermal conductivity and also their isotropic (aluminium) or anisotropic nature (graphite epoxy) (see Table 1.1). Consequently this study permits us to cover a broad spectrum of materials, especially since it is not possible to review all the possible materials. The illustrated cases are representative of major active TNDE possibilities and utilization of these techniques on other materials should not pose particular difficulty. Finally, it is important to note that even if illustration of one technique is studied for say, a graphite epoxy composite, it could be applied as well on an aluminium laminate and vice versa (of course with specific thermal behaviours). This is true for most of the techniques presented in this chapter.

4.2 Study of Graphite Epoxy Composites: Procedures, Investigation, Processing

As seen in Sect. 1.3, graphite epoxy structures are becoming more widely used because of their numerous qualities (Délouard et al. 1989). In this section we will see some TNDE techniques which can be applied to the inspection of graphite epoxy structures.

4.2.1 Impact damage

In the aerospace industry, the service life of a graphite epoxy component can be drastically reduced as a result of impact damage such as a bird strike (Craig and Chapman 1991). In fact, an impact of sufficient energy on such a composite can induce both fibre breaking and delamination. This damage can result from various accidents, such as a tool hitting the surface, or from combat conditions (such as by artillery). Generally, on the side where the impact takes place, there is only a barely visible scratch, while the opposite surface can be totally broken; this is the reason why such damage is often called *blind side* impact damage. An important point to note is that the opposite side of the component is often inaccessible once the complete structure is assembled. TNDE is thus well suited for *in service* damage detection in CFRP structures since it is able to detect shallow delaminations and since it can be employed directly on site, requiring only external access.

In order to illustrate TNDE deployment in context, tests of impact damage have been executed on Narmco 5217 graphite epoxy plates whose plies were stacked in various configurations. Impact damage was produced with a Dyna Tup drop weight impact tester. Figure 4.1 (see colour section) shows a typical result: large delaminated zones have been produced even at low energy impact (6 J) on unsupported thin samples (eight plies, 1.14 mm thick). In (a) the fibre orientation was $(0 \pm 45, 90)_s$ and in (b), $(0.90)_{2s}$. Samples were only supported by a foam pad. The imaged area is about 5×5 cm and the heating stimulation is about 15 °C above room temperature. This thermal stimulation took place with a 1 s pulse from one 1000 W projector (Fig. 3.6). In the thermograms, serious damage is visible on both sides of the impact area with the typical "butterfly" defect shape and a preferred orientation along the fibre of the first ply (from top to bottom in Fig. 4.1). In some instances, damage can extend deeply. For instance, in Fig. 4.1b, a wide horizontal delamination is visible. In the case of thicker supported graphite epoxy sheets, damage is of less importance.

4.2.2 Evaluation of Fibre Content in Graphite Epoxy Composites

Thermal methods are useful both for defect detection as seen in the previous section and also to characterize industrial materials (Cielo et al. 1986b). In this section we will present an example of such an investigation for the evaluation of fibre content in graphite epoxy composites.

The problem consists of evaluating the fibre–resin ratio in a graphite epoxy piece once it has been cured. Inappropriate temperature, pressure or assembly conditions can cause evacuation of a significant amount of epoxy resin from the composite. Areas having poor fibre–resin ratio present reduced mechanical properties.

A typical configuration is presented in Fig. 4.2. The workpiece is point heated with a laser beam while the thermal pattern is recorded with the infrared camera. If the material has an oriented structure with anisotropic properties such as in the case of a unidirectional graphite epoxy sheet, an elliptical pattern will be observed. The ratio (b/a) between the two ellipse axes is related to the square of the diffusivity ratio along longitudinal and transverse directions (Cielo et al. 1987b) and thus to the fibre orientation: a test on an isotropic material would give a circle instead of an ellipse. This property can be used to evaluate fibre orientation in extruded or moulded parts, and also to evaluate thermal conductivities of composite materials. This last approach is illustrated in Fig. 4.3 (see colour section). Thermograms (a) and (b) show the thermal pattern obtained after a 20 s heating using a 0.5 W argon laser on an eight-ply unidirectional oriented Narmco 5217 sheet cured (a) under high (690 kPa) pressure and (b) under reduced (69 kPa) pressure. The high pressure induces low resin content areas which have high thermal conductivity. For the rich epoxy content sample (Fig. 4.3b), the pattern is of smaller size and the central temperature is higher than for the low resin content pattern. Consequently the image difference (a) − (b) between the two patterns (Fig. 4.3c) is positive on the periphery and negative in the centre.

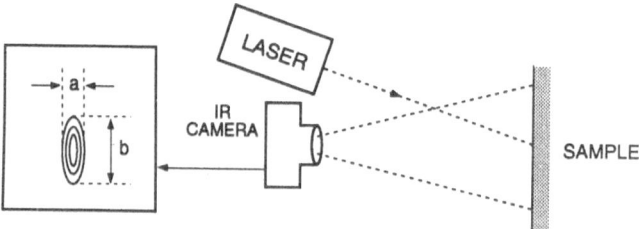

Fig. 4.2. Thermal analysis of fibre orientation in composite materials.

Such temperature inversion is typical of variation of thermal conductivity and is independent of the heating source and of the asorptivity or emissivity of the surface.

4.2.3 Delaminations

An appropriate model (Sect. 2.2) allows us to predict, up to a certain extent, the thermal behaviour of a particular workpiece under TNDE inspection, particularly when the analytical study becomes too complex. Modelling of graphite epoxy delaminations is a typical example of this situation (Fig. 4.4). This configuration is modelled in Fig. 4.5. In this figure, we notice the presence of the delamination does

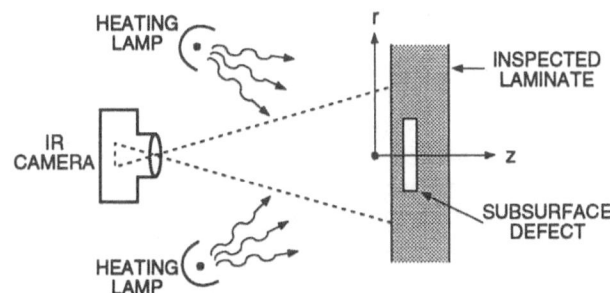

Fig. 4.4. Schematic diagram of the thermographic inspection configuration used for thermal modelling.

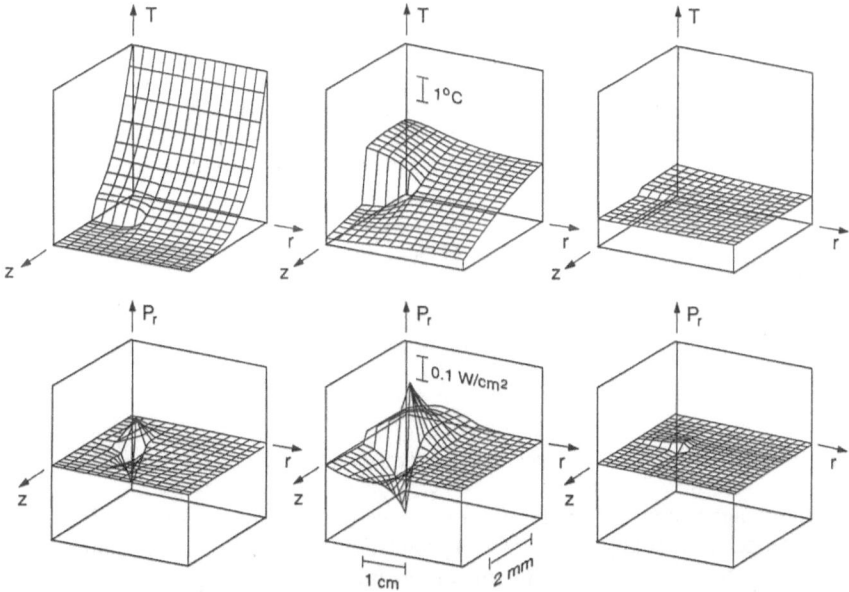

Fig. 4.5. Three-dimensional plots obtained from the model showing the computed temperature distribution (upper row) and the radial flux (bottom row) for the $r-z$ plane indicated in Fig. 4.4, respectively at 1.6 s (left), 6.3 s (centre) and 25 s (right) after beginning of heating.

not affect the surface temperature distribution at $t = 1.6$ s. After 6.3 s, a thermal contrast appears revealing the presence of the defect while after some 25 s, the radial thermal flow (bottom graphics in Fig. 4.5) is inverted: positive above the defect and negative below. At first it is located in a restricted volume around defect borders expanding to a larger volume with time. As a consequence, the thermal distribution just above the defect will be smoother on the surface ($z = 0$) with respect to the thermal distribution at the defect interface (top graphs in Fig. 4.5). This smoothing effect thus reduces the defect visibility and increases as the radius/depth ratio and the axial/radial thermal conductivities ratio reduce. These results can be confirmed experimentally.

In order to check for the limit of detectability of defects and illustrate the applicability of TNDE to CFRP components, tests were performed on graphite epoxy plates in which artificial known defects were inserted at the production stage. Such defects are made by inserting, at a known depth, before curing, two films of Teflon^{TM} 50 μm thick between two plies of prepreg. The two flms allow us to simulate a delamination rather than an inclusion.

Figure 4.6 (see colour section) shows thermograms obtained on samples in which implants of size (a) 20×20 mm, (b) 10×10 mm and (c) 3×3 mm were inserted 0.3 mm beneath the surface. Heating by reflection (Fig. 1.6a) was performed using the rosette of six projectors described in Section 3.1.3. Thermograms show the central portion of the heated area. The thermal pulse duration is fixed to 200 ms. Visibility of defects is adequate even if heating is restricted to a few degrees above room temperature. We must point out that in the case of the 3×3 mm defect, the heating pulse was a little longer in order to improve its visibility. This can be explained by the three-dimensional spreading of the thermal front (as noted above, Fig. 4.5), by the limited spatial resolution of the infrared camera (Sect. 3.1.1) and by the possible infiltration of epoxy along the edges of this small Teflon^{TM} implant.

These three-dimensional propagation effects were experimentally analysed (Fig. 4.6 c–e). In this case, Teflon^{TM} implants of 20×20 mm were inserted at depths of 0.3, 1.12 and 2.25 mm beneath the surface. The edges become more and more blurred as depth increases. The same situation occurs for the thermal contrast above the defect: it tends to vanish. This is a direct consequence of the thermal front propagation through diffusion. In this sense, defect visibility weakens when the thermal propagation defect–surface distance becomes similar to the propagation distance corresponding to the defect radius. This is why in an isotropic material, defects whose radius/depth ratio is less or close to the unity are difficult to detect. Graphite epoxy is an anisotropic material for which thermal diffusivity $\alpha_{\parallel}$ in a direction parallel to the fibres is typically ten times greater than the diffusivity $\alpha_{\perp}$ perpendicular to the fibres (Cielo 1984; Cielo et al. 1985b; Burleigh and De La Torre 1991). This thermal anisotropy increases the minimum ratio needed for the detection by a factor $\sqrt{\alpha_{\parallel}/\alpha_{\perp}}$ as we can also deduce from Eq. (1.1).

Spatial and Temporal Reference Thermogram Enhancement Technique

Special image processing techniques by spatial or temporal references are useful to improve defect visibility. In the visible spectrum, spatial reference techniques have already been used widely (see for instance Stansfield 1986). Spatial reference techniques are useful to compensate for repetitive variations in nonhomogeneity of the thermal perturbation source. This spatial reference technique consists of comparing the thermal behaviour of the specimen under examination with results obtained from a sound reference specimen (Monti and Mannara 1987). An example of this approach is illustrated in Fig. 4.8 (see colour section). Figure 4.7a shows a thermogram of Fig. 4.6e obtained above a 20×20 mm square defect, located 2.25 mm below the surface. In this image, the thermal contrast is poor, especially because of the nonhomogeneity of the heat source. These effects can be eliminated by subtraction from the reference image shown in Fig. 4.7b which was obtained in similar conditions above a sound area of the sample. This difference image (Fig. 4.7c) shows more distinctly the position of the defect. Notice the blurring effect around the edge in the subtracted image, due to the "large" defect depth in this highly anisotropic material.

When image degradation is nonrepetitive, for instance when surface emissivity of the workpiece changes in an unpredictable manner or when unknown thermal reflections are present, temporal methods of image processing can be applied to identify defects. An example of such an approach is depicted in Fig. 4.8 (see colour section). In this case, an artificial Teflon^{TM} defect of 10×10 mm was inserted, before curing, 0.5 mm under the surface of a graphite epoxy plate of 3 mm thickness. Images were recorded at (a) 3 s and (b) 5 s after starting heating. The thermal pulse was applied in transmission with a 1000 W projector placed on the back surface while the infrared camera was pointed on the front surface (Figs 1.6b and 3.6). During the experiment, strong thermal noise from a nearby welding station perturbed the measurement. The diagonal pattern visible on the images is caused by difference of reflectivity on the surface of this woven composite. Since this texture is the same on both images, reflection noise can be suppressed by subtraction of both images. This is shown in Fig. 4.8c. The subtracted image represents in fact the thermal evolution of the heated zone between the two observation times; it is thus related to thermal propagation properties of the sample.

In this section we have reviewed some basic and well established image processing techniques which allow us to improve visibility of relevant features in infrared images. One interesting aspect of these techniques is that, due to their simplicity, they can be implemented in real time. For the *spatial* reference technique this can be done on condition that we have recorded the reference image of the sound workpiece prior to the inspection, while for the *temporal* technique, subtraction can proceed just after the second image is acquired.

4.3 Study of Aluminium Laminates: Procedures, Investigation, Processing

4.3.1 General Considerations

TNDE, which is characterized by an excellent inspection rate, is an interesting alternative NDE technique in the aluminium industry where high production volumes are the rule. In this study, we will analyse the case of bonded laminates, aluminium–aluminium and aluminium–foam. They are increasingly used, for instance in the transport industry. As studied in Sect. 1.3, these laminates are made of thin aluminium sheets (a few millimetres thick) epoxy bonded either on a foam core or to another aluminium sheet.

Unpainted aluminium is characterized by a low emissivity ($\varepsilon \sim 0.05$). Consequently, thermal inspection will only be possible if high emissivity paint is first applied to the

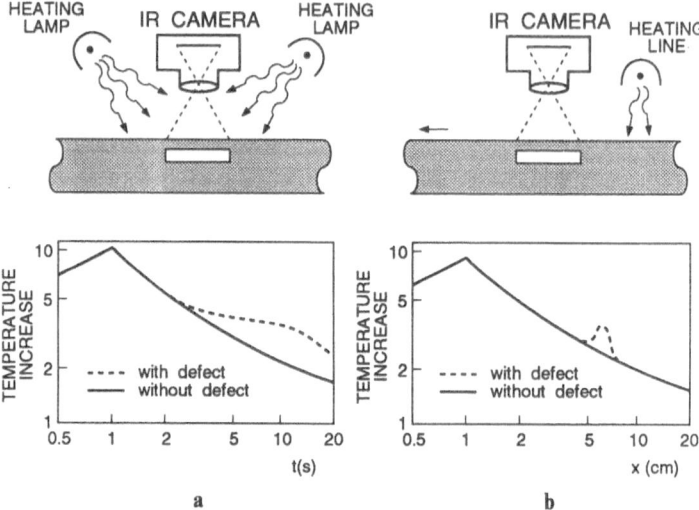

Fig. 4.9. Thermal inspection using **a** static and **b** mobile configuration. Shape of the obtained signals is also shown: **a** time pulse; **b** spatial pulse.

surface. This paint can serve as a primer coating or can be removed after the TNDE inspection, if required, by high pressure water jets. In the case of planar surfaces, we will review in Chap. 7 a method based on transfer imaging and which prevents such painting operations. Notice that in order to solve this low emissivity problem, all the aluminium samples tested in the book are covered with a high emissivity black paint ($\varepsilon \sim 0.9$).

In this section, we will illustrate two typical configurations (Fig. 4.9): the static configuration (time domain thermal pulse) in reflection or in transmission (Cielo et al. 1985b; Reynolds 1986) and the mobile configuration (space domain thermal pulse) in reflection or in transmission (Tretout and Marin 1985).

4.3.2 Static Configuration

As seen previously (Chap. 2), modelling is a valuable tool with which to predict defect visibility and select optimal inspection parameters. Figure 4.10 shows an analysis of the heat front thermal propagation in aluminium samples. The inspection geometry is shown in (a): the aluminium laminate is heated on one side using a rosette of projectors (two such projectors are pictured in Fig. 3.6). Surface temperature can be recorded either on the same side that the thermal perturbation is applied (i.e. in reflection) or on the opposite side (in transmission). A bonding defect consists of a lack of adhesive which can be modelled by a thin air layer of high thermal resistivity. The parameters for the modelling of Fig. 4.10 are as follows:

Thickness of aluminium sheets: 1 mm, thickness of the epoxy layer: 0.2 mm, radius of the defect: 1 cm.

Heat deposit: 2 W cm^{-2} for 1 s.

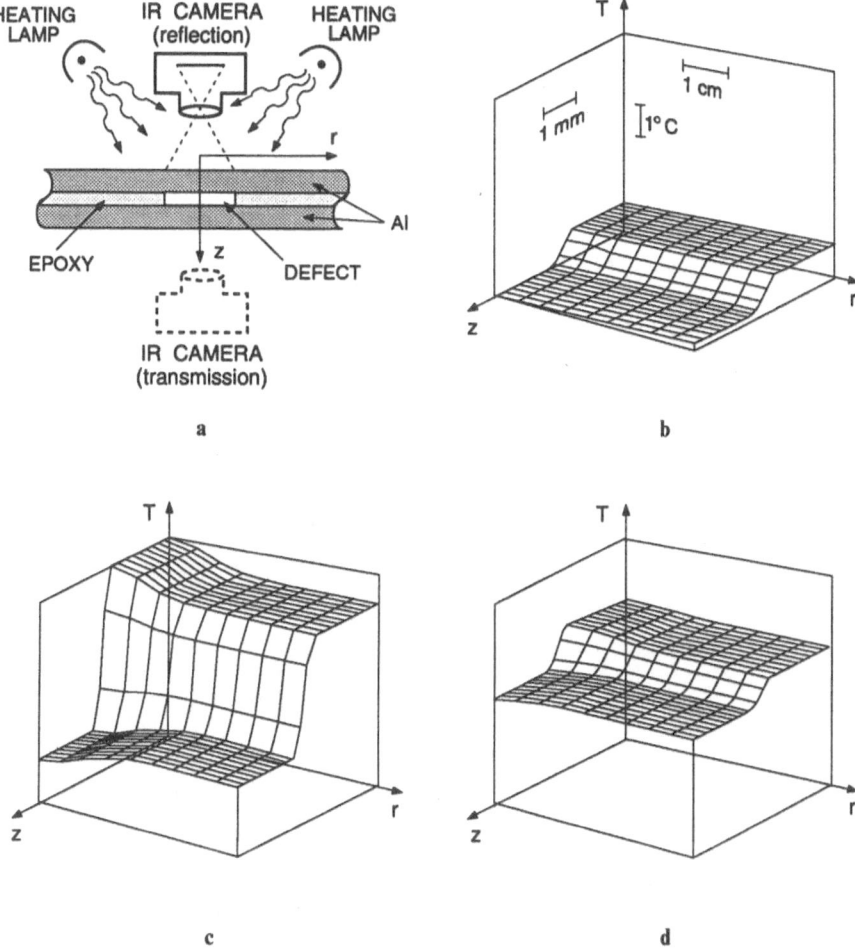

Fig. 4.10. Modelling of the inspection of an Al–epoxy–Al sandwich panel. **a** geometry; **b–d** thermal distribution computed 0.1 s, 1.6 s and 4 s after beginning of heating.

Thermal conductivities: 2 and 0.002 W cm^{-1} K^{-1}, specific heat: 1 and 1.5 J kg^{-1} K^{-1}, density: 2.7 and 1 g cm^{-3} respectively for aluminium sheets and epoxy layer.

Thermal resistivity of the interface: 83 cm^2 K W^{-1}; this corresponds to the thermal conduction through an air layer of 200 μm.

Losses through radiation: $5.67 \times 10^{-12}(T^4 - T_r^4)$ W cm^{-2} where T and T_r are the surface and room temperatures respectively (expressed in kelvins); convective surface losses: $10^3(T - T_r)$ W cm^{-2}.

Figure 4.10b–d displays temperature distribution in the $r-z$ plane pictured in Fig. 4.10a, respectively (a) 0.1, (b) 1.6 and (c) 4 s after starting heating. As we can notice, the temperature distribution is initially uniform over both surfaces thus preventing defect detection (Fig. 4.10b). The defect visibility increases up to a maximum (Fig. 4.10c) before vanishing through lateral diffusion (Fig. 4.10d).

The effect of three-dimensional heat flow spreading on defect visibility is particularly evident as shown in Fig. 4.11. For the three illustrated cases, the temperature difference between the centre of the delaminated area and the uniformly heated sound surrounding area is plotted for different specimen geometries. As we already mentioned, as a general rule, for a defect to be visible, the radius/depth ratio must be such that the radial flow around the defect is substantially smaller than the longitudinal flow through the bonded interface. For homogeneous materials, this condition limits the radius/depth ratio of a defect to values greater or close to unity since the limit of sensitivity for infrared radiometers is around 0.2 °C (Sect. 3.1.1). This is the case of Fig. 4.11a. In the case of bonded laminates (Fig. 4.11b), this ratio is close to 10 because of the small thermal conductivity of the bonding layer with respect to the high thermal conductivity of the aluminium skin. Greater values of radius/depth ratio are required

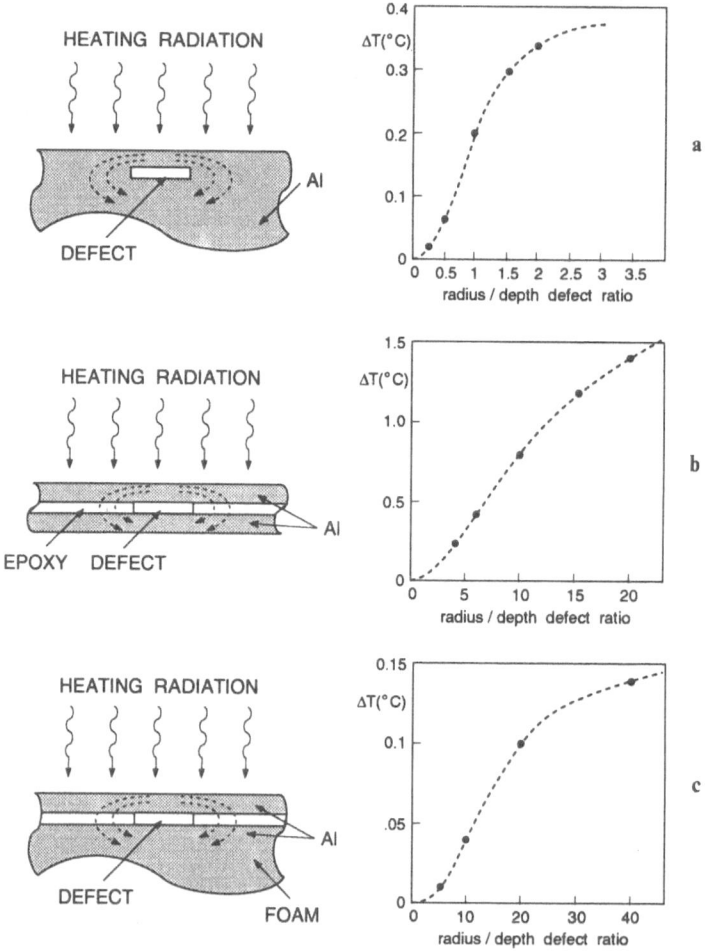

Fig. 4.11. Modelling of defect visibility: **a** homogeneous Al specimen; **b** Al–Al laminate; **c** Al–foam laminate. Thermal pulse lengths are 0.1, 1.0, 1.5 s and observation times are 0.16, 1.6, 2.7 s respectively for **a** to **c**. Heat deposition: 2 W cm^{-2} for 1 s.

in the case of aluminium and foam structures (Fig. 4.11c) because of the refractory properties of the foam (thermal conductivity $\sim 0.0004\,W\,cm^{-1}\,K^{-1}$, specific heat $2\,J\,g^{-1}\,K^{-1}$ and density $0.1\,g\,cm^{-3}$). This study shows, more formally than in the case of bonded aluminium laminates, TNDE is limited to the detection of large size shallow defects.

A configuration similar to the one of Fig. 4.10 has been used to inspect a panel constituted of two sheets of aluminium (of 1 mm thickness) epoxy-bonded on a polymethacrylimid foam core of 5 cm thickness. Artificial defects have been made by removing the epoxy adhesive film at specific locations before curing. Figure 4.12 (see colour section) shows a thermogram of the panel, 2.5 s after starting heating (fixed configuration in reflection, incident heating: $1\,W\,cm^{-2}$ for 1 s, field of view: 19×19 cm, triangular-shaped defect). The spatial subtraction technique (Sect. 4.2.3) is used here to suppress the nonuniform heating pattern (Fig. 4.13, see colour section). It is important to notice that the spatial subtraction technique (for which a defect free image recorded in the same conditions as for the inspected specimen is systematically subtracted from the images to be analysed) does not always bring such a dramatic improvement in defect visibility such as the one presented in Fig. 4.13.

Figures 4.14 and 4.15 (see colour section) present other investigations performed on aluminium laminates. Figure 4.14 shows bonding defect detection using the transmission configuration. In this case a flat bonded joint between two aluminium sheets of 1 mm thickness is inspected; a 1.5×1.5 cm area without adhesive is present. On the thermogram, the hot area on the right corresponds to the single aluminium sheet (one-sheet thickness) which heats more rapidly than the bonded area (two-sheet thickness). The heating must be quick to avoid lateral spreading of the thermal front from the hot plates to the bonded area. Special masks blocking the heating radiation from the lamp allow us to limit this disturbing phenomenon by stimulating only the joint area; this is the case of the left side of the bonding line visible in Fig. 4.14. This is especially useful for the repetitive inspection of identical samples.

In the case of components which are already at a higher temperature than room temperature because of various fabrication steps for instance, a cold thermal perturbing source can be used advantageously, as discussed in Chap. 1. An example of such analysis is presented in Fig. 4.15. The structure is initially at a temperature of some 10 °C above room temperature and a line of air jets is used to cool the inspected area quickly. The image shows the hotter central area which corresponds to the bonded area (oriented vertically on the thermogram). The cooler region at the centre of the bonding line reveals a bonding defect (lack of adhesive).

4.3.3 Mobile Configuration

The mobile configuration (Fig. 4.19b) allows high inspection rates to be reached. It is employed in reflection or in transmission. An interesting application concerns the inspection of flat joints in aluminium–aluminium laminates. The configuration in transmission is shown at the top of Fig. 4.16 (see colour section). The bonded joints are inspected by horizontal scanning of the panel, the joint being placed perpendicularly to the scanning direction. An 18 cm long 530 W line heating lamp serves for the stimulation. The vertical band in the central part of the image (Fig. 4.16, bottom) corresponds to the bonding line; it is cooler because of the greater material thickness of the workpiece in this area. In this image, a much cooler region corresponds to a bonding defect: the back surface thermal front takes more time to reach the front surface due to the increased thermal resistance of the subsurface defect it encounters on its way. The two aluminium plates have a thickness of 1 mm, while the overlapping zone is 2.7 cm in width.

An example of stimulation by propagation of a cool front in reflection is shown in Fig. 4.17 (see colour section). In this case, it is necessary to detect an unbonded area in an aluminium–foam laminate. The panel is first uniformly heated at a temperature of about 10 °C above room temperature. The surface is then cooled by a line of cool air jets during the lateral moving of the panel (at a constant speed of 2.4 cm s^{-1}). In the figure, two thermograms are shown, one for a sound area and one for an area where a circular-shaped debonding (4 cm diameter) is present. The defect is clearly visible at the image centre.

The same panel was tested in reflection using the line heating thermal source (Fig. 4.18, see colour section). The panel was scanned from right to left in front of the static line heater (here a 30 cm long 2500 W lamp), orientated in a perpendicular direction with respect to the figure plane. A polished back reflector was used to concentrate the thermal stimulation on the workpiece surface. This configuration allowed a 5 °C to be obtained on the stimulated surface while the panel was moving at a constant speed of 11.5 cm s^{-1}.

By comparison with Fig. 4.16 (or Fig. 4.19, see colour section), the fivefold less powerful 530 W lamp produces only a mere 1 °C contrast. This is a good illustration of the direct relationship between the temperature differential obtained and the amount of energy deposited on the stimulated surface (Eq. (2.24)). Powerful stimulation devices should be used wherever possible, but taking care not to damage the inspected surface through overheating.

It is also interesting to compare the radiative heat injection (Fig. 4.18) and convection heat removal approach (Fig. 4.17). A reduced thermal contrast is obtained with the cool air stimulation due to both the reduced heat capacity of air (see Sect. 5.3.2) and the smaller temperature differential between the thermal perturbing source and the inspected surface (case of air versus high temperature radiative source). Recall of Eqs (2.25) and (2.26) is also of interest to understand such differences: radiation stimulation is a function of the temperature differential to the fourth power, while convection is directly proportional to the temperature differential.

Lateral Scanning Defect Detection Algorithm

The last application allows us to introduce an algorithm especially suited to defect extraction in images obtained in the mobile configuration where workpieces are laterally scanned. The principle is as follows. Observation of the image sequence during the motion of the panel in front of the heating lamp reveals a very interesting fact: even in the presence of noise, defects are easily seen because of their lateral motion. Noise is random while defects follow panel motion. On the contrary, if only one image is observed, this dynamic effect is lost and the distinction between defects, noise and sound areas is less distinct. This is because the eye–brain entity performs an integration of the dynamic noise, while it cannot proceed in such a manner when the noise pattern is fixed (Gauffre 1986). This phenomenon can also be observed by watching an image sequence recorded from a low-cost VCR and comparing the perceived quality obtained with a particular image seen while the VCR is in "pause" mode: the fixed image is very poor, while the image sequence is acceptable. The effect is used profitably in some low-rate image transmission systems where dynamic noise is superimposed with images in order to have them more appealing to the viewer by reducing the *contour effect* due to the quantification performed on a limited number of grey levels (Hall 1979).

The algorithm is based on the analysis of many images (two typically) recorded at different moments. Notice that the use of motion is described in the literature as a means to improve detection of features in images, see for instance the works of Jain et al. 1979; Zisk and Wittels 1987, p. 3; Boivin et al. 1989.

In this algorithm (originally proposed by Maldague et al. (1988), a subtraction by spatial reference is first performed as described previously (Fig. 4.19a–c). Subtracted images are next binarized (Fig. 4.19d, e) through application of a threshold function. Binary images are coded with only two levels: high and low. Notice that for this application what is of interest is the defect detection and in particular, defect localization and gross size estimation; consequently relative intensity of defects is not relevant in this study.

In a first step, knowing the lateral speed of the panels (expressed in pixel s^{-1}), it is possible to compare two (or more) images by taking into account their lateral motion. Each binary image is divided into a succession of small matrixes (3×3). The matrix of one image located in (a, b) is compared with the corresponding matrix in the other image (recorded a little time later) and which is located at the position $(a + \Delta a, b)$ where Δa is the interval due to the lateral motion (along the x axis). The passage rule to obtain the result is simple: if at least one element is at high level in both considered matrixes, the result matrix (output image O_1) is set to high, otherwise, it is set to low. This simple technique allows us to reduce significantly the amount of noise present in the image (Fig. 4.19f).

This first step is, however, not sufficient to *purify* image O_1 sufficiently. For this reason another noise suppressing technique is applied, this time by working only on image O_1. At this stage, a longer matrix is used. Its size corresponds roughly to the minimum defect size we wish to detect. In our example, this corresponds to a matrix of 32×32 pixels. This large size matrix is scanned over image O_1 from left to right and top to bottom, at every location and the ratio of the number of high level pixels to the number of low level pixels is computed. If this ratio is smaller than a pre-established value (20%), all the considered windows in O_1 are set to low level, otherwise the window content stays unaffected. This kind of processing, by *erosion*, is quite radical and permits us to eliminate completely the residual noise present in O_1; the only structures subsisting in the new output image O_2 are the defects, if any (Fig. 4.19g).

The erosion process is generally followed by a third stage, *dilatation*, in order to recover somewhat the original shape of the defects which went through two successive erosive phases. The third stage is similar to the second stage since it works also with a matrix scanned over the image, as before. However, this time, the matrix size is fixed at 5×5 pixels, a sufficiently small size to track defect edges. The rule is as follows: if at least one element of the considered window is at the high level, the whole window is set to high, otherwise nothing happens. This dilatation method has the advantage of filling the "holes" in the defects and to reconstruct partly their edges; the third image output O_3 is shown in Fig. 4.19h.

As a possible improvement to the algorithm, a flexible matrix grid would be more appropriate than a fixed grid in order to avoid edge truncation. Drawbacks of this modification are increased programming complexity and slower execution time, however. The algorithm was revealed to be very powerful for the segmentation of thermograms recorded during lateral inspection; it requires two input parameters:

the threshold (to binarize the images) and also the lateral motion speed (expressed in pixels s^{-1}). In the case of repetitive inspections, these values are set in advance. It is, however, important to check periodically the validity of these parameters. In fact, the use of an absolute threshold is not always reliable. In the next section, we will introduce another algorithm which allows us to relax this constraint.

Before concluding this section, it is interesting to point out that even if binary images have been used in the algorithm, if we perform an "AND" operation between output image O_3 and the first analysed image, original grey levels of the defects are restored, thus enabling quantitative analysis of the detected defects (Chap. 6).

In this section we have presented a series of investigations, procedures and image processing techniques relevant to TNDE in order to detect and improve the visibility of defects. Although the analysis was applied to the particular case of aluminium laminates, these concepts are, of course, not restricted to this particular kind of workpiece.

4.4 Automatic Defect Detection

4.4.1 General Considerations

In the previous sections of this chapter, we presented investigation procedures and image processing techniques useful for thermographic infrared inspection in the mobile or static configuration, either in reflection or in transmission, with a heat injection or heat removal approach. We reviewed the detectability limit of TNDE based on the minimum radius/depth ratio. We also reviewed methods to improve defect visibility on TNDE images. In this section, we will discuss defect detection algorithms.

In the case of repetitive TNDE inspection, for instance on the production line, we are interested in the automatic processing of the recorded thermograms. In this section we will present an algorithm for the automatic detection and extraction of defects observed on thermograms. The goal is to produce a complete map of the inspected component where both defect locations and gross shapes are depicted. Since a list of defects with their approximated size is available, automatic inspection becomes possible: the components inspected can be sorted on an accepted/rejected criterion based on a probability detection curve (Fig. 1.7). Obviously, the inspection method must be able to identify all the critical defects with reliability, this means the detection critical threshold must be greater than the threshold for false readings. Rejected components must be eliminated from the production line or fixed if the cost of refurbishing is acceptable. In this case, a more sensitive NDE technique such as ultrasound can be used to assess defective zones with greater resolution.

For instance, in the case of bonded aluminium–foam laminates we studied previously (Fig. 4.11c), we noticed that the threshold for detection ($\sim 0.2\,^{\circ}$C) corresponds to a radius/depth ratio of the order of 40. Detected defects (Figs 4.17 and 4.19) have a ratio of this order. This is at the limit of detectability for TNDE. It is important to notice that, in the case of these aluminium–foam components, defects which must be absolutely detected, in "real life", have a much greater ratio. Typical defect sizes are of the order of 30×30 cm ($\sim 1\,\mathrm{ft}^2$) with a corresponding radius/depth ratio of about 150. An undetected defect of this size can be very dangerous due to the degradation in mechanical properties it will cause. TNDE is thus particularly well suited as an inspection technique for this kind of material.

Defect Detection Algorithms

Concerning detection algorithms, the Society of Photo-Interpretive Engineers (SPIE) conference series on image processing algorithms is a good source of information on this subject (see e.g. Pennington and Moorhead 1990) as also are the Institute of Electrical and Electronics Engineers (IEEE) Transactions on Pattern Analysis and Machine Intelligence (PAMI) and Systems, Man and Cybernetics (SMC). Many other pertinent journals are mentioned in the reference section of this book.

There exists a wide variety of algorithms which purposes are to perform *image segmentation*, that is to separate regions of interest in images. For instance, we can cite the work of: Green (1970); Zucker (1976); Schachter et al. (1979); Minor and Sklonsky (1981); Lineberry (1982); Eshera et al. (1986); Elliott et al. (1986); Zisk and Wittels 1987; Zheng and Basart (1987); Williams (1988); Krishnan and Walters (1988); Baker et al. (1988); Stein and Heller (1989) (method based on fractals for crack detection); Rodriguez and Mitchell (1989) (decomposition of the image in rectangular regions to extract local background). Techniques based on edge detection (such as the work of Marr and Hildreth 1980; Laurendeau and Poussart 1985; Chen and Medioni 1989), regions growing around key points called seeds, histogram analysis (this is a widespread technique since it does not require complex computations; see for instance the work of Kapur et al. 1985) or symbolic modelling have been published and utilize either discontinuity or similarity of characteristic attributes (such as pixel intensity, gradients, ...) in order to label all the pixels within the image, that is in order to associate them with a particular region. Considering the wide variety of image analysis problems, the variety of algorithms is extreme. Generally algorithms are ad hoc and if applied in another context, they can fail pitifully (Fu and Mui 1981). The main reason behind this is because an image represents an enormous number of possibilities: if we consider for instance the case of the infrared images produced by the thermographic inspection station studied as an example in this book, even with their relatively small size, 68 (Maxrow) $\times$ 105 (Maxcol): 7140 pixels on 8 bits, we already have $(2^8)^{7140}$ possibilities!

Another important aspect is the image background. In TNDE, because of heating effects (nonuniformity), image borders tend generally to be hotter than the other parts of the images. This has inspired some authors to develop *trend removal* elimination procedures (Zheng and Basart 1988; Burch 1987; Doering and Basart 1988). The basic idea consists of producing a *synthetic image* from the original image by making use of a polynomial function in order to obtain, after subtraction or division of the synthetic image, a uniform background against which defects appear more clearly. Next, a threshold segmentation algorithm based on valley detection in the image histogram (see e.g. Chow and Kaneko 1972) is used to identify a threshold between the two modes (defects and background) at either the global (whole image) or local level (using a small running window over the image). However, this method does not work very well if the separation between the two modes is not sharp (see Rounds and Sutty (1980), Sahoo et al. (1988) and Abutaleb (1989) for a survey of threshold techniques). Although the trend removal approach is well suited for some types of images such as radiographic images, it proves deceptive in TNDE images (Fig. 4.20, see colour section).

The use of transform approaches such as the Fast Fourier Transform (FFT) or the Fast Hartley Transform (FHT), although very effective, is not really practical on the production line because of the great computing power associated with these techniques (e.g. Rosenfeld 1969; Bracewell 1984; Mitra et al. 1988). Other detection algorithms have been reported in the case of military target detection. Burton and

Benning (1981), in their study, introduce the notion of *pixels on target* (POT); in conjunction with SNR evaluation; this helps to compare algorithm performance.

Validity of Defect Detection Procedures

A difficult question is to determine whether or not the segmentation is valid. An approach to the question is to consider the degree of agreement with the human interpretation (Zahn 1971). In fact the eye–brain combination is extraordinarily powerful, considering the fact that only about 50% of the cortex cells are dedicated to the vision task.

This is why, in most cases, an experienced operator is perfectly able to segment TNDE images as they appear originally. In fact, it seems the human nervous system analyses images using threshold techniques which allow objects to be separated on the basis of their relative intensity (Bell 1987, 1988). Obviously *cultural* experience is also necessary in performing such a task. This is why, for instance, a newborn baby has to learn all about its surrounding environment before being able to recognize objects. This cultural aspect corresponds to *heuristic* rules in the case of machine vision algorithms. This discussion shows that image transformation is not always necessary to perform artefact detection in NDE images, not counting the additional processing time needed to convert images into the transform domain (and back). Finally, if comparisons are to be made with the human vision, it is also important to realize that the eye response is logarithmic which allows it to have an extremely large dynamic range.

4.4.2 Image Formation

We will now review a technique which is useful for image formation. In order to obtain a high inspection rate, the amount of computation to be performed on the images obtained from the inspection station should be restricted to a minimum. For this reason, the number of images to analyse after the inspection is performed must be restricted as much as possible. Obviously, this is in contradiction with the requirement to record all the thermal history curve of the inspected part after stimulation by the thermal perturbation in order to catch abnormal transient thermal events (Fig. 1.2).

As seen previously, images can be obtained from either a static or mobile configuration. In the static configuration (Fig. 4.9a), the camera observes the same surface continuously and a succession of images is obtained beginning at the instant $t = 0$ corresponding to the firing of the thermal pulse (e.g. the time when heating lamps are turned on). Notice that in some instances, it may be of interest to start image acquisition slightly before the thermal pulse in order to obtain a reference "cold" image which will allow the spurious effects of thermal reflections to be reduced by subtraction from other images (Sect. 4.2.3 and Chap. 6).

For the static configuration, the *moment method* initially proposed by Balageas et al. (1987b) can be used. The temporal moment of order M for temperature T_0 on a sound area is defined by:

$$M = \int_0^\infty \Delta T_0(t)\, dt \tag{4.1}$$

The Δ operator means we are interested in an increase of temperature with respect to ambient room temperature T_a: $\Delta T = T_0 - T_a$. This moment M tends to infinity. If a defect is present, we can form $[\Delta T_D(t) - \Delta T_0(t)]$ where T_D corresponds to the temperature above the defect zone. Consequently, we can evaluate the temporal moment ΔM of order zero. It can be demonstrated that this moment has a finite value equal to:

$$\Delta M = \int_0^\infty [\Delta T_D(t) - \Delta T_0(t)]\mathrm{d}t = \frac{Q}{R_{def}}\left(1 - \frac{z_{def}}{L^2}\right) \qquad (4.2)$$

where Q corresponds to the absorbed energy by the sample of thickness L while R_{def} is the thermal resistance of the defect and z_{def} is its depth. If the sample is very thick, ΔM becomes equal to Q/R_{def} and defects will appear with the same contrast whatever their depth. This relationship (4.2) can be applied simply by adding together all the recorded images in the time domain. This also has the advantage of improving the SNR by the square of the number of summed images, as we have discussed previously. Consequently, if this summation process is applied, it is not necessary to use the noise reduction techniques described in Sect. 3.2.2. However, Bosher et al. (1988) mention that care must be taken in this summation process: not "all" the images must be added since, for a given thermal event, the thermal contrast tends to vanish as images in which it is absent are summed together.

This summation method can be applied in order to reduce the number of images to process following the TNDE inspection procedure. After (or during) the experiment, it is only necessary to add together all the acquired images in a given time window. Care must be taken when selecting this window to have it sufficiently wide so that thermal contrasts of potentially present defects have had the opportunity to develop sufficiently, taking into account the thermal diffusivity α of the analysed material (Eq. (2.1)). However, in practice, this constraint is quite flexible. If we call the image obtained in this fashion I during time window $[t_a, t_b]$ corresponding to individual images $G(t_i)$, we have

$$I = \sum_{t_a}^{t_b} G(t_i) \qquad (4.3)$$

In the case of the mobile configuration, where the infrared camera records the complete motion of the inspected component, this direct summation process cannot be directly applied since the field of view is constantly changing. In this case a special technique can be used in which specific columns of pixels are extracted from every recorded image in order to *reconstruct* the whole component as seen at a particular time. Next, the reconstructed images are summed together following the method of Eq. (4.3) in order to obtain the I' image (the prime indicates a reconstructed image).

In Fig. 4.21, we illustrate the principle of inspection in the case of components in motion in the field of view of width L and at time $t_0, t_1, \dots, t_i, \dots, t_N$. This corresponds to acquisition of N images during the inspection experiment. In Fig. 4.22, we show the reconstruction process for the image G'_k obtained through juxtaposition (operator J) of columns $C_k(t_i)$ extracted from images $t_0, t_1, \dots, t_i, \dots, t_N$:

$$G'_k = \int_{i=0}^N C_k(t_i) \qquad (4.4)$$

with

$k = 0,1,2,\dots,\text{Maxcol-1}$

$\text{Maxcol} = \text{number of columns in one image}$

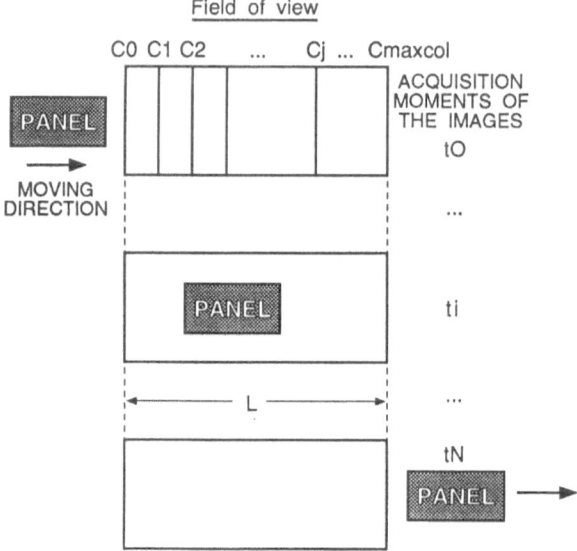

Fig. 4.21. Motion of the panel in the field of view (mobile configuration in reflection).

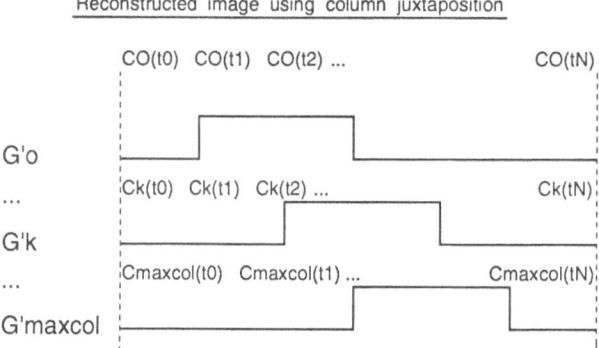

Fig. 4.22. Reconstruction process for the images.

Due to the motion of the component within the field of view, all the reconstructed images G'_k correspond to the total observation of the component when it is in the thermal state $t(k)$ since the component is observed at the same distance from the heating unit (same extracted column k in the field of view) and since the lateral speed v is constant, from Fig. 4.23, we obtain

$$t(k) = \frac{L'}{v} + \frac{LC_k}{vC_{\text{Maxcol}}} = \frac{1}{v}\left[L' + \frac{LC_k}{C_{\text{Maxcol}}}\right] \tag{4.5}$$

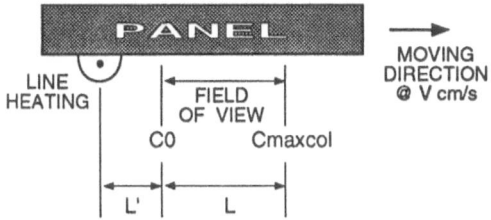

Fig. 4.23. Studied geometry (see text).

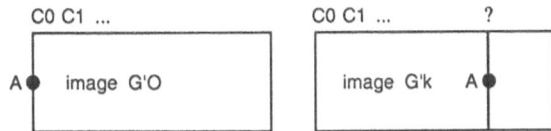

Fig. 4.24. Study of the shifting in the reconstructed images (see text).

From this, we see that images G' correspond somewhat to images G of Eq. (4.3) since

$$
\begin{array}{llll}
\text{Image} & \text{Corresponds to component} & N_{col} \\
& \text{at time} \\
G'_0 & \left[\dfrac{1}{v}\right]L' & 0 \\[2mm]
G'_k & \left[\dfrac{1}{v}\right](L' + L(C_k/C_{Maxcol})) & k & (4.6) \\[2mm]
G'_{Maxcol-1} & \left[\dfrac{1}{v}\right](L' + L) & Maxcol-1
\end{array}
$$

Where N_{col} is the column number of the original infrared images having Maxcol columns.

In Fig. 4.22, the horizontal scale is a temporal scale directly related to acquisition time of images $t_0, t_1, \ldots, t_i, \ldots, t_N$, and every reconstructed image corresponds to a particular thermal state of the component (Eqs (4.5), (4.6)). However, we notice in this figure that the component is not present at the same position in the sequence of reconstructed image $G'_0, \ldots, G'_{Maxcol}$. This shifting of the image G'_k obtained by *juxtaposition* of the kth column of the Nth acquired image can be evaluated. The question is as follows: In which column of image G'_k will the point "A", which appears in column 0 of image G'_0, appear? This situation is depicted in Fig. 4.24. Since the point is moving at speed v, it will move from column C_0 to C_k within the field of view in a time interval given by (Fig. 4.25).

$$
t = \frac{L}{v}\frac{C_k}{C_{Maxcol}} \tag{4.7}
$$

We obtain N images in time t_N; one image is thus acquired in t_N/N s. Consequently the point A will appear in the "column"; notice that here we talk about the columns of the reconstructed images which correspond to image number from 0 to N:

$$
\text{Sh\_col}(k) = \frac{L}{v}\frac{C_k}{C_{Maxcol}} \bigg/ \frac{t_N}{N} = \frac{L}{v}\frac{NC_k}{t_N C_{Maxcol}} = \frac{L}{v}\frac{RC_k}{C_{Maxcol}} \tag{4.8}
$$

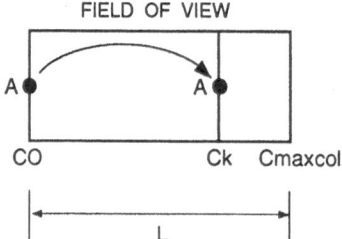

Fig. 4.25. Study of the shifting in the reconstructed images (see text).

where R is the acquisition rate (number of images acquired per second). Every G_k' image will thus have to be shifted by Sh_col(k) columns before proceeding to the summation following Eq. (4.3) as in the static case:

$$
\begin{aligned}
I' &= G_0' + \text{SHIFT}[G_1', \text{Sh\_col}(1)] + \cdots \\
&\quad + \text{SHIFT}[G_i', \text{Sh\_col}(i)] + \cdots \\
&\quad + \text{SHIFT}[G_{\text{Maxcol}}', \text{Sh\_col(Maxcol)}]
\end{aligned}
$$

$$
I' = \sum_{i=0}^{\text{Maxcol}} \text{SHIFT}[G_i', \text{Sh\_col}(i)] \tag{4.9}
$$

where $\text{SHIFT}[G_i', \text{Sh\_col}(i)]$ corresponds to the shifting operation of Sh_col(i) columns on image G_i'.

This study shows it is possible to obtain a *reconstructed* image corresponding to the whole width of the inspected component: due to the lateral motion of the component, its full width is inspected. It is important to point out that this study (computation in Eq. (4.9)) is, however, an approximation, since we take into account the *apparent separation* of the image columns instead of the *real separation* which takes into consideration the SRF (as mentioned in Sect. 4.1.1). However, as we will see later, this procedure proves adequate, at least for the defect detection analysis.

Since the temporal information is lost in the image formation process (of static (Eq. (4.3)) and dynamic (Eq. (4.9)) configurations), the depth of detected defects cannot be computed (unless it is known due to the geometry of the inspected components such as in the case of known depths in bonded laminates). This is also the case for defect size: no *corrective factor* can be computed to recover the real shape from the apparent size (Chap. 6). However, these limitations do not restrict automatic defect detection as mentioned before.

We studied in Chap. 3 what corrections are needed to apply to raw images for temperature conversion. In the case of the static configuration for which the whole temperature history curve of detected defects is available, after the defect detection step, it is only needed to reprocess and correct the images to obtain a quantitative information as we will see in Chap. 6. In the case of the mobile configuration, this study can also be done on the basis of individual reconstructed $\text{SHIFT}[G_k', \text{Sh\_col}(k)]$ images (Eq. (4.9)). However, due to the approximate nature of the reconstruction process, the margin of error in such quantitative computations may be unacceptable.

In order to maximize the computational speed and taking into account the previous limitations on defect size and depth, the automatic detection algorithm can be applied directly on raw thermal value images; this saves the time needed to execute the temperature computation step. The fact the algorithm works even in the absence of these corrections is a positive aspect. Obviously, the algorithm described in the next sections can be applied to temperature-converted images; the main difference between defect detection on raw and on corrected temperature images will be the time of execution (needed for temperature computation).

Moreover, with the same desire to maximize execution speed, images I and I' can be converted in *unsigned character* type which takes less space in computer memory. In this way one pixel is coded on one byte in computer memory instead of four or eight bytes required for the coding of the floating point type of variables. This allows us also to make use of *integer arithmetic*, which is faster than *floating point* arithmetic in most computer implementations. The conversion of an image in which pixels are expressed in floating point values (I_{float}) to an image in which pixels are expressed in characters (I_{char}) is performed as follows:

$$I_{char}(i, j) = m I_{float}(i, j) + b \qquad (4.10)$$

where

$$m = \frac{B_{max} - B_{min}}{F_{max} - F_{min}}$$

$$b = B_{max} - m F_{max}$$

B_{max}, B_{min} are maximum and minimum values of image I_{char} (in 8 bit implementations, 255 and 0 respectively)

F_{max}, F_{min} are maximum and minimum values found in image I_{float}

i, j are all the pixel positions in the images

All the subsequent image processing steps will be performed on image I_{char} or I'_{char} which will be denoted by I for simplicity.

In the following examples, the thermal perturbation source deposits energy on the specimens and the inspection proceeds in reflection; consequently, potential defects will be represented in image I by areas of higher temperature with respect to their immediate surroundings. Defect edges will be represented by ramps of temperature which span over a few pixels because of the three-dimensional spreading of the heat flow, as explained previously. Because of this ramp aspect, the use of edge detector operators is less attractive.

4.4.3 Automatic Segmentation Algorithm

Part I: Defect Localization

It is interesting to determine both the location and the gross size of defects which may be present in TNDE images. In Chap. 6, we will present an approach for the quantitative evaluation of defects once they have been detected; the detection procedure we will discuss now is nevertheless informative. For instance, in the case of aluminium bonded laminates previously studied (Sect. 4.3), defect depth is known since it is given by the thickness of the aluminium sheet above the bonding epoxy layer. Moreover, in many situations, the important point is whether or not there is a defect present in a given part.

The algorithm we will review (initially proposed by Maldague et al. (1990b)) is based on the fact that TNDE images have a limited content of spatial features. This is a very different situation with respect to visible images characterized by complex edge structures. Moreover, this algorithm makes use of some heuristics (Tou and Gonzalez 1974, p 18; Bell 1987; Bell et al. 1988) which are not so different from the expert system approaches exploited by some researchers (Nazif and Levine 1984; Gilmore 1985; Stansfield 1986).

In a first step, we proceed to defect localization and then, specific thresholds are found in the image. The originality of this algorithm comes from the fact the seeds are first reliably located based on a global sorting process within the image (a *seed* is defined in the literature as the centre point of a detected defect, the hottest for hot thermal perturbation schemes). One threshold will be established per detected defect. Each threshold is found by means of a region growing approach which starts at the central point of a detected defect and stops when either an image border is hit or the number of pixels agglomerated around the seed increases abruptly (meaning the *image background* is reached).

The first part of the algorithm, once completed, produces, for each defect found in image I, the location of the hottest point (i.e. highest grey level pixel, or seed). In order to limit the computation, we suppose that defects have at least one pixel at an intensity greater than the image average A (Zahn 1971; Mao and Strickland 1988). In this respect, many background pixels will be neglected since only pixels greater than A will be processed. Of course all the pixels of the image can be processed, but the computational time will be greater. This procedure is a short cut rather than a limitation; it allows the processing time to be approximately halved.

All the pixels $I(i, j)$ of the image are compared with A; if $I(i, j) > A$, they are charged in a four-vector structure (first initialized to zero):

$\mathbf{vx}(r) = j$ i.e. position of pixel $I(i, j)$ along the column

$\mathbf{vy}(r) = i$ i.e. position of pixel $I(i, j)$ along the row

$\mathbf{gl}(r) = I(i, j)$ i.e. grey level of pixel $I(i, j)$

$\mathbf{lb}(r) = $ label associated with pixel $I(i, j)$

where r is the rank in the vector structure: $0, 1, 2, \ldots, [\frac{1}{2}(\text{Maxcol} \times \text{Maxrow}) - 1]$.

Initially vector $\mathbf{lb}$ is set to zero so that no label is associated to any pixel. Next, all the non-zero elements of vector $\mathbf{gl}$ are sorted in decreasing order. For all r values, we obtain

$$\mathbf{gl}(r) \geqslant \mathbf{gl}(r+1) \tag{4.11}$$

The content of the two other vectors $\mathbf{vx}$ and $\mathbf{vy}$ is also processed to avoid mixing the information related to individual pixels. After this *grey level sorting*, sorting is performed spatially so that, for every position r of pixels having the same grey level value, we have

$$\mathbf{vx}(r) \geqslant \mathbf{vx}(r+1) \tag{4.12}$$

As mentioned previously, the content of the other vectors is reorganized to avoid mixing pixel information. These sorting operations ensure that neighbouring pixels belonging to the same class (see below) will be close in the vector structure. Once the two sorting procedures have been accomplished, the absolute grey level values of pixels (the $\mathbf{gl}$ values) are no longer important besides the relative rank of the pixels with respect to the others in the vector structure. In this sense, we can say that the algorithm adapts itself to the histogram distribution of image I.

The last step consists of labelling all the pixels, that is to assign them to a given class. Class 1 is assigned to the first pixel of the structure ($r = 0$). All the subsequent pixels are tested with all the previously labelled pixels starting with the last that was tested (i.e. its closest neighbour). This speeds up the labelling process (the likelihood of having neighbouring pixels of the same class being high). The criterion for assigning a new label to a pixel r_i is:

$$\text{if } \mathbf{vx}(r_i) - \mathbf{vx}(r) > \text{MND}$$
$$\text{or } \text{ if } \mathbf{vy}(r_i) - \mathbf{vy}(r) > \text{MND} \tag{4.13}$$

The constant MND is called *Minimum Neighbour Distance* and it is established through trial and error. Since MND represents a distance between pixels rather than an absolute grey level value, it is independent of the image; rather it depends on the field of view size and on the minimum defect size to detect (Fig. 1.7).

This first part of the algorithm can be understood more easily with respect to Fig. 4.26 where the one-dimension case is shown. Two defects (blobs) A and B are represented. Suppose grey level value of the hottest pixel of blob B is denoted by $B(x_B)$ with x_B referring to the x position along one image row and $A(x)$ is the grey level of any pixel of A. The algorithm will proceed assigning "label A" to analysed pixels sorted in decreasing grey level values up to the moment it has to label pixel $B(x_B)$. At this point, $B(x_B) = A(x_A)$ and a different label will be assigned to blob B ("label B") at the condition that: $d > \text{MND}$ (where $d = x_B - x_A$), independently of absolute intensity of blobs A and B. In the case of an image I, this process is extended to both axes (rows and columns).

The labelling process stops whenever all the pixels present in the vector structure are labelled or that a predefined number of labels have been posted (that is, a predefined number of defects has been found in the image I). In fact as pixels of low grey level values are labelled, the risk of false readings grows. Moreover, if the condition to reject an inspected part is that at least a predefined number of defects is detected, the computational time can be reduced if the labelling process is stopped as soon as this predefined number of detected defects is reached (in the case of a bad component). In order to validate this procedure, the procedure can be redone with

$$\text{MND}' \leftarrow \text{MND} + 1 \tag{4.14}$$

until, for two successive tests, the same defect set is obtained. Each testing is rapidly executed since sorting within the vector structure is already available. It is only necessary to reinitialize the label vector **lb** to zero and to redo the labelling process with MND'.

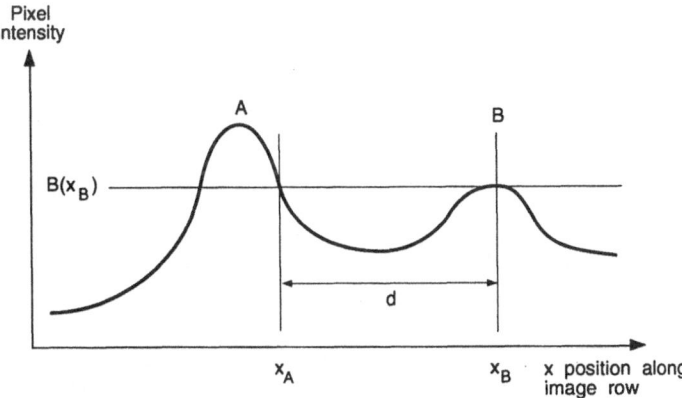

Fig. 4.26. Principle of the algorithm for defect detection (see text).

Part II: Defect Edge Estimation

At this stage of the algorithm, all the seeds corresponding to the defects are available. In this section, gross defect shape will be estimated by growing a region around the seeds (besides region growing, other techniques for boundary finding have been reported, see for instance Deriche (1990) Shann and Oakley (1990). Each seed will be processed individually and one threshold per defect will be established (Green 1970). If many defects are present in an image, many thresholds (one per defect) will be obtained. The purpose of the technique discussed in this section is to give a rapid estimation of defect shape; in Sect. 6.3.4 a method will be discussed to determine defect shape quantitatively.

In image I, for the defect (i.e. the seed) located in (i_d, j_d), the threshold Th is first set:

$$Th = Th_{max} = I(i_d, j_d) \qquad (4.15)$$

where $I(i_d, j_d)$ = grey level of pixel located in (i_d, j_d) in image I, and the number of neighbours $Nr(Th_{max})$ having the same grey level value as $I(i_d, j_d)$ is computed using a recursive procedure and assuming an eight connectivity (Horn 1986, p 66). The search is then redone with

$$Th' \leftarrow Th - 1 \qquad (4.16)$$

until, using a recursive procedure, an image border is hit. At this moment, the vector **Nr** holds, for all the potential thresholds Th, the number of pixels agglomerated around the seed located in (i_d, j_d). Figure 4.27 shows a schematic example of the **Nr** content (Fig. 4.28 shows typical real values). If present in **Nr**, a sudden increase in the number of agglomerated pixels indicates the image background intensity is reached. This is especially the case of reconstructed images I' in the mobile configuration. Using simple heuristic rules, this threshold is located at Th_{est} if the correspond-

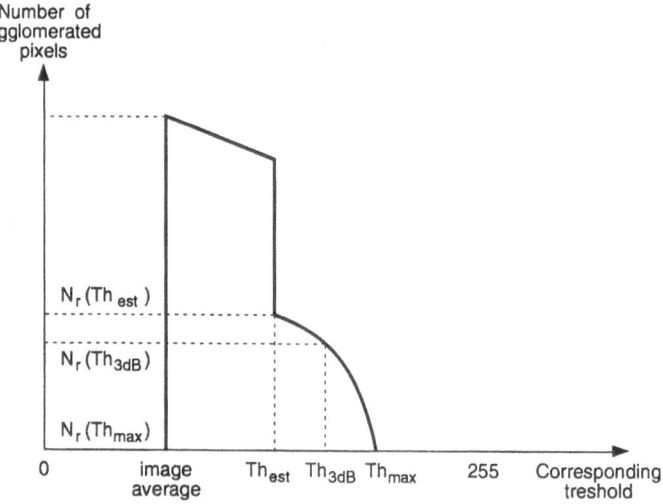

Fig. 4.27. Schematic variation of the number of pixels agglomerated around a defect (seed) as a function of the threshold. The sudden increase corresponds to the background value, thus allowing us to establish the threshold Th_{3dB}. Note that computation was done only for pixels with grey level above image average (see text); this explains the abrupt transition seen on left for lower grey level values.

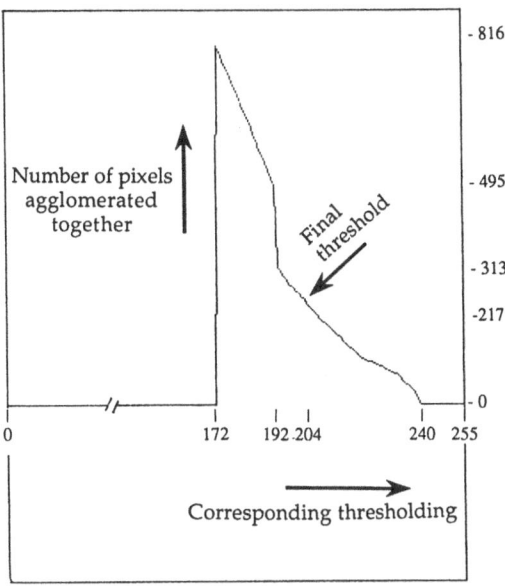

Fig. 4.28. Typical example of vector **Nr** content, case of the central defect of image AL.1 of Fig. 4.31b.

ing difference $|\mathbf{Nr}(\mathrm{Th}_{est}) - \mathbf{Nr}(\mathrm{Th}_{est} - 1)|$ is found greater than three times the average step computed from the differences $|\mathbf{Nr}(\mathrm{Th}_i) - \mathbf{Nr}(\mathrm{Th}_{i-1})|$ between adjacent positions in vector **Nr** from $i = \mathrm{Th}_{max}$ to 0. In fact, this corresponds to computing the derivative of vector **Nr** and apply rules to locate and validate the sharp background transition. Notice that Part I of the algorithm could be applied on vector **Nr** derivative to locate the sudden increase of agglomerated pixels which corresponds to a large derivative value.

If no sharp transition is found within **Nr**, the maximum is kept:

$$\mathrm{Th}_{est} = \mathrm{Th}_{max} \qquad (4.17)$$

Many tests indicated that the threshold Th_{est} obtained using this procedure does not correspond perfectly with manual segmentation. A corrective 3 dB factor allows us to obtain a better agreement. The defect threshold is thus set at

$$\mathrm{Th}_{3dB} \Leftrightarrow \frac{1}{\sqrt{2}} \mathbf{Nr}(\mathrm{Th}_{est}) \qquad (4.18)$$

The threshold specified by Eq. (4.18) was found adequate in most cases. This use of *heuristic* rules can be understood since it gives to the interpretation program knowledge not available otherwise.

4.4.4 Results and Discussion

In this section, we will present some typical results obtained with the segmentation algorithm described in the previous sections and with images obtained on a variety of samples tested either in the static or mobile configuration. The mobile configuration is of interest since it allows large surfaces to be inspected rapidly.

The segmentation algorithm allows a *map* of the inspected surface to be obtained where positions of the detected defects are indicated. If only defect detection is necessary, Part II of the algorithm is not required. It may, however, be of interest to evaluate the severity of the damaged areas in order to reject false readings based on probability criteria of size, of relative intensity, or of other characteristic features (perimeter, surface, etc.; see Vavilov 1980; Bow 1984, p 69; Girard and Algazi 1985). Detection of false readings is related to the notion of a *detection probability curve* studied previously, in particular the question of small defects taking into account the amount of noise present in the thermograms.

Generally, an interesting approach to validating segmentation and results of NDE techniques consists of establishing a statistical data base using for instance the Tanimoto detection criterion which can be defined as follows in inspection situations (Yanisov and Yanisova 1984):

$$R = \frac{N_R - N_M}{N_R + N_F} \qquad\qquad (4.19)$$

where

N_R = number of real defects

N_M = number of missed defects

N_F = number of false readings

Obviously, to use this statistical method, a great number of components must the considered and destroyed after the inspection in order to determine N_R, N_M and N_F. If component destruction is not practical, it may be valuable to go through production and maintenance records in order to establish the Tanimoto criterion value (Kraft and Wing 1981).

Figure 4.29 (see colour section) shows some segmentation results for a static configuration in reflection. The images are organized in pairs as follows: on the left raw images, on the right segmented images. The components tested are typical of the aerospace industry.

In the mobile configuration, the field of view depends on the size of the inspected component as well as on the moving speed; a smaller Minimum Neighbour Distance is selected in order to pick up smaller defects which are less visible since they fill less space in the reconstructed images. Figure 4.30 (see colour section) shows some results; in all cases, MND = 9. It is worthwhile mentioning that in these images (except b) the broken texture is due to the low image transfer rate of the equipment used (3.8 images per second). Figure 4.30b was obtained with image transfer performed from an experiment recorded on a VCR tape. Since more images are available in this case (30 images per second), the texture of the reconstructed image is smoother.

As stated previously, temporal information is lost in both the *reconstruction* and in the *summation* process (Eqs (4.3) and (4.9)). Consequently, it is not possible, at this stage, to evaluate quantitatively defect depth and size. This was not the purpose of this study. It is important to notice, however, that, for this reason, identical defects located at two different depths will appear in the segmented image as having different diameters.

Returning to the Tanimoto criterion explained previously, the performance of this segmentation algorithm was checked and yields to a 90% figure. Of course, unless a component is cut into pieces after the inspection, it is difficult to estimate the quality of the inspection/detection/interpretation procedure, especially in the case of "real defects". For instance, in Fig. 4.31 (see colour section), we show the image obtained by static inspection of an impact-damaged graphite epoxy component; two defects show up in the segmented image, and the butterfly shape previously discussed (Sect. 4.2.1) is easily distinguished. In order to determine whether the second defect is discrete, or only an outgrowth of the main defect (this is likely to be the case for impact damages), the part would have to be cut into pieces and/or inspected with another NDE technique such as water-immersed ultrasonic C-scan.

This segmentation algorithm is well suited to automatic defect extraction in low spatial content images such as in the case of TNDE images, while the defect maps produced by the algorithm are easily interpreted. The algorithm uses information at the global level to ease the decision making processes at the *local* level. Since no complex operation is required, it can be easily implemented. This is an important aspect in the case of real-time inspection/interpretation procedures where hardware implementation becomes mandatory to achieve high throughput (Wallace 1988; Wolfe 1988). Parallel implementation is also possible, especially with Single Instruction Multiple Data (SIMD) machines (Nickolls 1990). For instance, considering multiple processor elements (PEs) working simultaneously, defect shape estimation can be performed on a one PE per defect basis. Expected throughput improvement is interesting: processing time will be the same whatever the number of defects.

The MND constant based on spatial distance rather than on absolute intensity proves robust and relatively insensitive to noise, background uniformity, defect size, provided that defects are not too small, of course (Weszka et al. 1974; Bell 1987; Eshera et al. 1986; Mao and Strickland 1988), number of defects (Elliot et al. 1986), defect intensity (Schachter et al. 1979) and orientation (Richardson and Schafer 1987).

In Chap. 6, we will see how the quantitative analysis completes results presented in this chapter. Note, however, it is estimated than the detection–localization approach suffices in 80% of inspection situations (Barre 1988).

The concept of automatic inspection through image reconstruction and segmentation is well suited to repetitive inspection of large components whose surfaces are relatively planar (nonplanar surfaces will be discussed in Chap. 10).

For cases where the inspection must be performed manually by an operator, it is important to maximize defect visibility in the thermograms. Techniques reviewed in Sects 4.2 and 4.3 are pertinent to these tasks.

Conclusion. In this chapter we have studied methods and image processing for the internal stimulation scheme.

Chapter 5

Internal Thermal Stimulation: Methods and Image Processing

5.1 General Considerations

In the previous chapters, we explained how infrared thermography can be employed for the inspection of materials and structures by means of thermal transient perturbation. More precisely, we reviewed in Chap. 4 many different methods and image processing techniques useful for the TNDE inspection task. In Chap. 4, the thermal stimulation was accomplished by means of lamps, thermal radiators, lasers, air/water jets, etc. In all cases, an *external* thermal stimulation was performed (with respect to the workpiece). We saw that thermal contrasts obtained over defective areas were generally small, of the order of a few degrees only since the thermal stimulation heated (cooled) the surface under inspection by 10 °C at most (e.g. Fig. 4.11). As discussed in Sect. 1.5, one of the limitations of TNDE is precisely the difficulty to obtain a uniform, short, high intensity energy deposition over a wide surface with the added constraint of not causing damage.

There exist situations where access is available from inside the structure being inspected. In these situations, the structure can be inspected in transmission (Fig. 1.6b) rather than in *reflection* (as in most cases presented in Chap. 4). Moreover, instead of using a thermal perturbation source of a radiative type such as a lamp or a radiator, the thermal perturbation can be accomplished, if possible, by changing the temperature of a fluid circulating inside the structure. Depending on the heat capacity of the fluid, large temperature differences can be employed (within the acceptable limit to avoid component damage) in order to produce high thermal contrast, leading to reliable thermal inspection.

In this chapter, we will illustrate the internal stimulation technique by reviewing two typical applications which use common fluids: liquid (water) and gas (air). More specifically, these case studies concern evaluation of corrosion damage in pipes due to the flow of corrosive liquids and analysis of internal structure of jet turbine blades. The case studies presented will permit us to introduce various experimental methods as well as image processing techniques.

The basic principles of TNDE in the case of *internal* thermal stimulation have been proposed by the following authors among others: Beynon (1982); Reynolds and Wells (1984); Ding (1985); Bantel et al. (1986); Reynolds (1986); Maldague and Dufour (1989); Maldague et al. (1990a,c).

5.2 Case Study I: Evaluation of Corrosion Damage to Pipes

It is estimated than in industrialized countries, losses caused by corrosion and wear count for around 3% of GNP. In plants where industrial processes require circulation of corrosive fluids, the corrosion problem is thus an important issue.

In fact, circulation of corrosive fluids can damage pipe walls considerably because of *local erosion*. This phenomenon is more prominent in pipe bends where turbulence and vortexes often take place, increasing wear on the surface significantly. The phenomenon takes place in three stages (Kvernes et al. 1988): a long period of time where the surface is attacked; an active period where damage is no longer superficial (the core starts to be damaged and cracks can appear due to fatigue); *catastrophic failure* when the thinned wall cannot support the internal pressure any longer and actually breaks down. Wall thinning must be taken into consideraiton in order to avoid accidents which may happen if corroded walls explode under internal pressure, especially in nuclear or petrochemical industries.

Traditionally, industry relies on the statistical replacement of pipe sections judged potentially dangerous after a certain period of time: this method may prove costly if intact sections are replaced. Other methods are used such as visual inspection performed with boroscopes introduced directly inside pipe sections (Svedemar 1985; Kobayashi and Ueda 1987). Such a method is not always practical because of the dismantling involved.

Other NDE techniques (especially ultrasonic and X-rays) are good candidates for this task. Some problems of accessibility and of surface curvature slow down point-by-point ultrasonic operation, while security aspects of X-rays may restrict its deployment. For this application, TNDE offers many advantages (see Sect. 1.5): no contact; surface scanning (rapidity); no harmful radiation involved; ease of deployment.

It can be demonstrated that, under stationary conditions, for a pipe in which a fluid is circulating, temperature differences between zones of different thickness are very small. A much more important contrast can be obtained in transient conditions. If the surface of a thick piece is heated starting at $t = 0$ with a source delivering W watts per surface unity, temperature difference at depth z below the surface follows the expression (Carlslaw and Jaeger 1959, Chap. 10, Eq. 9)

$$T - T_a = 2W \left\{ \sqrt{\frac{t}{\pi K \rho C}} \exp\left(\frac{-z^2}{4\alpha t}\right) - \frac{z}{2K} \operatorname{erfc}\left(\frac{z}{2\sqrt{\alpha t}}\right) \right\} \tag{5.1}$$

where K is the thermal conductivity, ρ the density, c the specific heat and $\alpha = K/\rho c$ the thermal diffusivity of the material, and T_a is the temperature before the perturbation. Because of the abrupt spatial distribution of the erfc$(\cdots)$ function, the argument of the function becomes prominent in thermal transient conditions so that after a time t (starting from the beginning of heating), the thermal front will reach a depth z_{th}:

$$z_{th} \approx \sqrt{\alpha t} \qquad (5.2)$$

This relation was seen in Eq. (1.1). It indicates that for a pipe section whose wall thickness has been reduced by a factor of two due to the corrosion, the thermal perturbation will reach the exterior surface after a period of time four times shorter than for an intact uncorroded wall section. After a time (counted from the beginning of the thermal perturbation) of the order of z_c^2/α, where z_c is the thickness of the corroded wall, this wall section will have reached a different temperature with respect to nearby sound sections submitted to the same perturbation and still at the temperature T_a (temperature before the perturbation). Consequently, if measurements of the propagation time are performed, the wall thickness can be evaluated.

In order to demonstrate these concepts, a simple method can be used. A transient thermal perturbation is obtained inside a pipe section by changing the temperature of the circulating fluid. Observation of the temperature distribution on the external surface is performed with the infrared camera. Pipe sections having a thinner wall will have their temperature affected first.

We study now an application of this method in the case of a corroded 90° bend section of stainless steel pipe. The experimental set-up is shown in Fig. 5.1. The original wall thickness was 5 mm. The pipe section was connected to two supplies of cold (6 °C) and warm (40 °C) water with a flow of nearly $120 \, l \, min^{-1}$. Thermal inspection proceeded as follows: flow circulated in the pipe until a steady thermal regime was reached on the external surface; next the flow temperature was changed and a sequence of infrared images was recorded during the thermal transition. High flow and large temperature differences yield excellent thermal contrast.

Different configurations were tested: cool or warm transition; forward or reverse flow. The same thermal signatures were obtained in all cases. All these experimental conditions revealed the same abnormal temperature pattern which corresponds to corroded areas. This last fact was checked after a visual examination was performed

Fig. 5.1. Experimental apparatus with infrared camera visible on left as well as the pipe bend section connected to the cold and hot water supply taps (top right).

on the corroded pipe section cut into two pieces over its length. Figure 5.2 (see colour section) shows a typical thermal transition from hot to cold. The corroded areas are visible from $t = 2.17$ s to $t = 3.54$ s. The still-hot areas present at $t = 16.6$ s correspond to a sound area four times thicker than the damaged area, thus explaining the long delay required for the cold front to reach the outside surface at these locations. As noted before, under the steady regime (Fig. 5.2a–f), no temperature difference was observed and thus no damage could be detected. Notice that vortexes, responsible for corrosion in specific spots due to increased water flow at these locations, amplify the thermal transient effect observed by the infrared camera on the outer surface. Figure 5.3 (see colour section) shows the image recorded at $t = 2.72$ s during the cold-to-warm transition, the same pattern as for the warm-to-cold transition (Fig. 5.2c) is observed.

The thermal transmission method is fast and easy to deploy if variation of the temperature of the circulating fluid is possible and does not require extended dismantling of the pipe. Moreover, since the thermal perturbation source is located inside the pipe, it does not generate any thermal noise. This is an advantage with respect to radiative sources of thermal stimulation (e.g. Fig. 4.8, colour section).

In this section, a TNDE method based on an internal thermal perturbation source has been reviewed and applied to the inspection of corroded pipe sections. This internal thermal perturbation method allows us to obtain large thermal contrasts due to the large temperature differential of the stimulation source. This TNDE method can be applied to other hollow structures as we will see in the next section.

5.3 Case Study II : Inspection of Jet Turbine Blades

In this section, we will illustrate another possible deployment of the internal perturbation technique, for the inspection of the inner structure of jet engine turbine blades. In this study we will use two different thermal stimulation media: a gas rig (air) and a liquid rig (water).

Following thermodynamic laws, engine efficiency depends on the temperature differential of the thermodynamic cycle. Jet engines are thus designed to operate at high temperature, of the order of 550 to 1000 °C, and metal protection is ensured through the use of thermal barriers and appropriate cooling methods. The turbine blades must be continuously cooled through internal circulation of a fluid in order to sustain the extreme operating temperatures present in the engine; this is especially important for the blades exposed to combustion gases.

Considering the high rotation speed of the blades during operation, failure of one blade can have dramatic consequences (Bedrossian and Slazas 1987). In order to prevent these kinds of accidents, the temperature of the blades can be continuously monitored, in service, through a fibre optic pyrometer. This allows early detection of a sudden abnormal blade temperature rise before blade failure (Ding 1985; Florin 1987). It is also important to check blade integrity at the engine manufacturing stage. In particular, it is important to ensure that blade cooling ducts are unblocked (Reynolds 1986).

Several nondestructive evaluation methods have been proposed and are used for blade inspection tasks, for instance techniques using ultrasound (Gostelow et al. 1986), neutron radiography (Dance and Middlebrook 1978), X-ray radiography

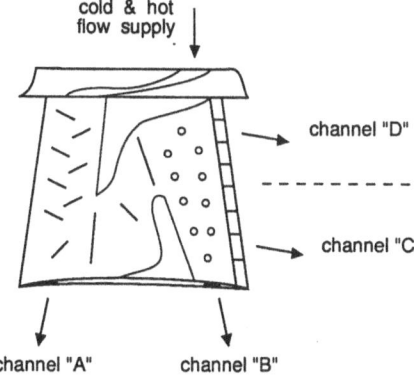

Fig. 5.4. Simplified schematic diagram of the turbine blades tested showing the various ejection channels on the blade edges (pressure side).

(Oliver et al. 1986), computerized tomography (Tonner and Tosello 1986; Morisseau et al. 1987) or manual investigation. Manual testing is performed by inserting thin wires through blade channels to check open passages. This method is labour intensive and cannot be automated.

The internal perturbation technique described in Sect. 5.2 can be applied for the evaluation of internal structures of jet engine blades. Figure 5.4 shows a simplified schematic of the internal structure of the turbine blades tested; note the various ejection channels on the blade edges. Defective blades can have these ejection channels blocked by tiny solidified metal debris following injection moulding during the blade manufacturing process.

Flow circulation of a liquid (water) or a gas (air) at different temperatures allows, in the transient thermal regime as we discussed previously, detection of blocked passages by means of the delayed arrival of the thermal front. In other words, if a portion of the blade wall thickness is increased by a factor of two because of machining errors or other causes, the thermal disturbance will reach the outer surface of this portion after a time period four times longer than the surrounding unblocked portions. This is the basic principle of the method; areas of the blade where the cooling flow is more efficient (unblocked) will show greater outer surface temperature fluctuations after a time period t of the order of $t \sim z_{th}^2/\alpha$, following Eq. (5.2), where z_{th} is the blade wall thickness.

Water is a good medium for thermographic inspection because of its good cooling properties and consequently, it is often used to investigate thermally blades or other hollow structures. Conversely, air offers poor thermal removal efficiency. For a simplified one-dimensional stationary analysis, we can write for the thermal exchange between the flow medium and the blade structure, considering only convection (Thomas 1980, Eq. 1-28):

$$q_c = hA_S(T_S - T_F) \qquad (5.3)$$

where q_c is the rate of heat transferred from the surface at uniform temperature T_S to the fluid of temperature T_F, A_S is the surface area and h is the coefficient of heat transfer. With the reduced thermal capacity of air, in order to obtain a similar thermal contrast with respect to water stimulation, the temperature differential $(T_S - T_F)$ for an air rig should be greater than for a water rig and the flow rate through the blade must be more important. Consider for instance that if we compute (Thomas 1980, p 374) the coefficient h for a circular structure, 1 cm in diameter and 2 cm in length, we obtain $h = 96\,\mathrm{W\,m^{-2}\,^{\circ}C^{-1}}$ for air at a rate of $10\,\mathrm{m\,s^{-1}}$ and $h = 350\,\mathrm{W\,m^{-2}\,^{\circ}C^{-1}}$ for water at a rate of only $1\,\mathrm{cm\,s^{-1}}$ (a flow 1000 times smaller). As a result of this simplified analysis, we expect fewer visible time-delay effects with air stimulation than with water stimulation.

5.3.1 Experimental Analysis

These concepts can be applied as follows. A thermal transient is generated inside the blade under inspection by changing the temperature of the circulating fluid (water or air) and then observing the transient temperature distribution on the outside blade surface with an infrared camera (pressure or suction side). Areas with thinner walls

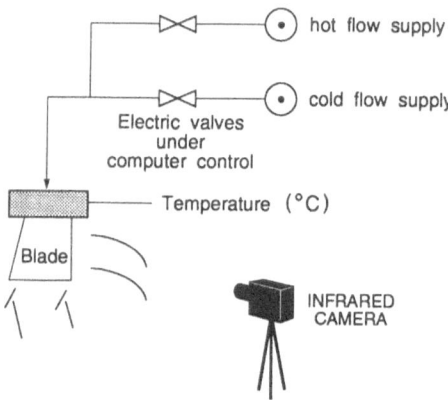

Fig. 5.5. Schematic diagram and photograph of the experimental apparatus for the thermographic inspection of a turbine blade using internal stimulation (transmission approach). Hot and cold water supplies are visible; they are directly controlled by the computer using electric valves. This enables good reproducibility of measurements.

will have their temperature affected first, blocked areas will be affected later: the thermal signature of the blade will depend on the flow circulation inside the blade. Any blocked passage will thus produce a different thermal response leading to a possible detection and recognition of which particular channel is blocked. The basic arrangement is shown in Fig. 5.5. Cool and hot flow lines serve for the stimulation of the blades, and two computer-controlled electric valves allow either the cool or hot flow to circulate through the blade. For the air rig, a room temperature air feed line can be used for the cool line, while the hot line can be obtained by diverting the flow of the room temperature air tap through a serpentine held in an oven. For the water rig, two temperature-controlled recirculating baths can supply cool and warm water.

Thanks to the repetitive computer-controlled thermal stimulation, several thermal transitions (or "runs") hot to cold (or the reverse) can be repeated successively with the images of each run averaged together to increase the SNR (Ahmed et al. 1987; Kuo et al. 1989).

For every thermal transition, two images are gathered for each run. This allows us to implement the time subtraction technique of Sect. 4.2: two images obtained at different times are subtracted in order to remove the unwanted reflections and background (considered constant at the two acquisition times). The first image is recorded during the steady warm regime while the second image is recorded during the thermal transient (from warm to cool). The opposite scheme, where the first image is recorded during the steady cool regime and the second image recorded during the cold to warm thermal transient yields the same kind of thermal signature (as stated in Sect. 5.2). The subtracted image shows more clearly the thermal signature of the blade. This subtraction scheme is particularly helpful if unpainted blades are inspected (see Figs 5.9 and 5.10 colour section). In this case the low emissivity ε of about 0.3 for a nickel alloy material makes the blade surface act as a mirror (reflection coefficient: $1 - \varepsilon$) and unwanted thermal reflections caused by warm surrounding bodies superimpose on the thermal emission of the blade itself, making the analysis harder, if not impossible. In order to increase surface emissivity, blades can be covered with a paint of high emissivity in the spectral region of interest before the thermographic inspection is undertaken. This painting step, although convenient in a research investigation, is totally inappropriate on the plant floor where no foreign material is allowed in engine components; the subtraction technique is thus essential under these conditions.

Typical results obtained for the five categories of blades are shown (see colour section) in Fig. 5.6 (air stimulation) and Fig. 5.7 (water stimulation): very characteristic thermal signatures depending on which of the cooling passages is blocked are obtained. In the case of air stimulation of Fig. 5.6, differences are particularly clear if profiles along a bottom row and an edge column are drawn (Fig. 5.8).

Figures 5.9 (air stimulation) and 5.10 (water stimulation) (see colour section) show the result of the inspection in the case of painted and unpainted blades. In both cases the subtracted images reveal a similar thermal signature to that for painted blades (the uneven emissivity distribution over the unpainted surface in Fig. 5.10 is due to the black lettering used for blade marking at production stages). When inspecting unpainted surfaces, it is important to note that the time subtraction approach will work only if the unwanted thermal reflections are the same for the two acquisition times.

Figures 5.10 (water stimulation) and 5.11 (air stimulation) (see colour section) show how averaging many consecutive runs together helps to increase the definition of the thermal signature by limiting the noise content (Ahmed et al. 1989). We recall that the SNR is increased by the square root of the number of summed runs (Shepherd

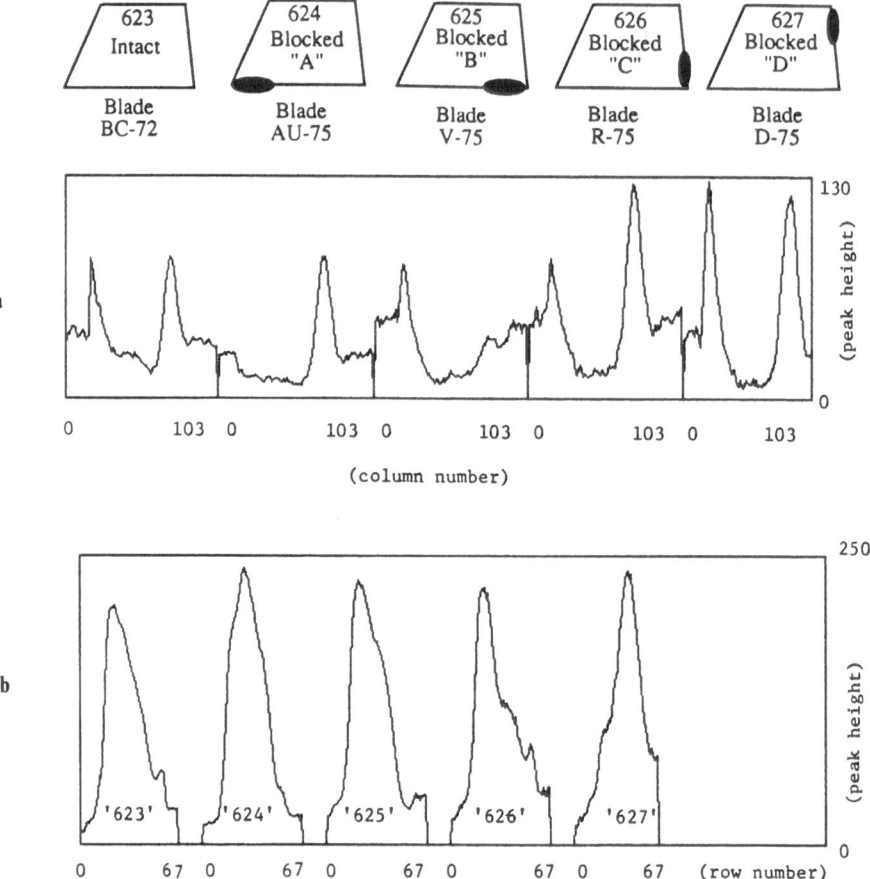

Fig. 5.8. Air rig. **a** Profiles of one row at the bottom of the subtracted images of Fig. 5.6 (row 53). The peaks are related to the presence of open cooling channels at the bottom of the blades. **b** Profiles of one column at the blade edge of the subtracted image of Fig. 5.6 (column 77). The position and width are related to the presence of open cooling channels at the edge of the blades.

and Moorey 1987). However, even taken individually (no averaging), the thermograms offer the same basic geometric organization. Notice thermograms shown in Figs 5.6, 5.7 and 5.9 are the result of averaging many consecutive runs together.

5.3.2 Thermal Signature Analysis

The detection algorithm presented in Sect. 4.4 can be used to discriminate automatically between the different blade signatures obtained in the previous section: the idea is to be able to recognize automatically which channel is blocked, if any, in a given thermogram (e.g. front channel "A" blocked, rear channel "B" blocked). We recall that this algorithm detects blobs in an image; examination of Figs 5.6 and 5.7 reveals that location and amplitude of blobs in the subtracted image (Fig. 5.7) or in profiles extracted from the subtracted image along the bottom row and edge column (Fig. 5.6) are dependent on the type of blade which is tested. Determination of blob location and amplitude leads to possible detection and recognition.

Water Rig

In the case of the water rig, many different tests (78 in total) were carried out with the five categories of blades (one intact and four with artificial defects of types A to D, Fig. 5.4). For all the "runs" of a given category, the computer program plots a drawing showing the sketch of the blade as well as the positions where peaks are detected within the image (Fig. 5.12). We recall that one "run" consists of one test of a blade: acquisition of two images during the thermal transition warm to cold, subtraction of these two images and detection of the high intensity peaks by the algorithm in the subtracted image. Due to the close proximity of features of interest present within the images, the MND parameter is set to 5 in all cases (Eq. (4.13)).

The four boxes in Fig. 5.12 correspond to the various possible blocked channels. It can be seen that, depending on which of the blade channels is blocked, more peaks will be located in the other boxes (with the exception of type D defect, Fig. 5.12) making the interpretation very simple: if no peak is detected within a box, this means the corresponding blade opening is blocked (defect). The reason behind this can be understood as follows: if a channel is blocked, the water flow will divert towards the other open channels. Figure 5.12 is plotted in black and white so that the colour-coded intensities of the detected peaks (intensity of the brightest pixel within the peak, i.e. seed intensity) are missing, thus preventing visualization of detected peaks located on box borders. This is, for instance, the case of the few peaks present in the box of type D defect (Fig. 5.12) which can be rejected with a simple validation criterion based on a minimum intensity of the detected peaks present in a given box. Using the algorithm of Sect. 4.4 and this "box intensity criterion", a reliability factor of 95% (Tanimoto criterion, Eq. (4.19)) is obtained.

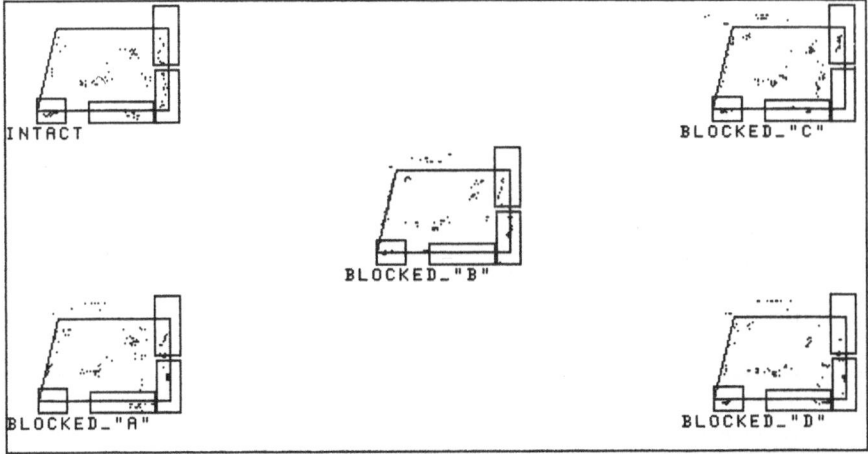

Fig. 5.12. Water rig: automatic interpretation of the inspection procedure (see text).

Air Rig

For the air rig, it was found more convenient to operate on profiles (bottom row and edge column) extracted from the subtracted image of a "run". A typical investigation using the algorithm of Sect. 4.4 is shown in Fig. 5.13. The correct behaviour of the algorithm, independently of the absolute values, is particularly clear in Fig. 5.13a, where the two main modes of the profile are correctly located even under noisy conditions. In all cases, the MND parameter (Eq. (4.13)) is set to 20; a large value is retained this time since it is necessary to extract low spatial frequency peaks from the profiles.

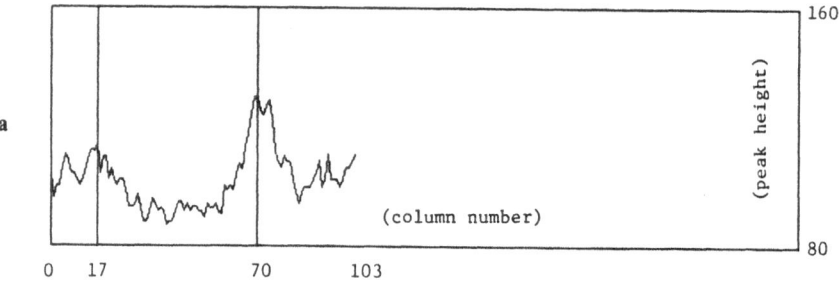

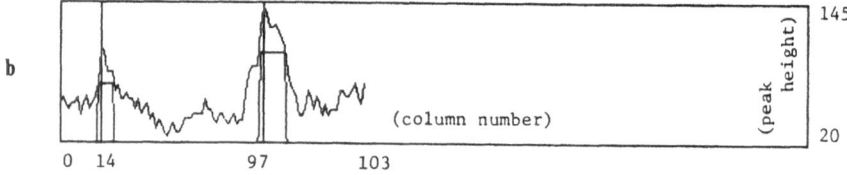

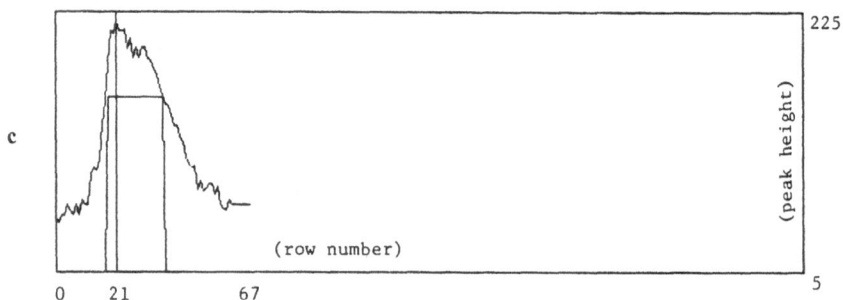

Fig. 5.13. Investigation by the one-dimensional-peak finder algorithm. **a** Detection along row 53 (run 5 of experiment 626, Fig. 5.6). **b** Detection and peak width (at 3 dB) along row 53 (run 11 of experiment 626, Fig. 5.6). **c** Detection and peak width (at 3 dB) along column 77 (run 5 of experiment 623, Fig. 5.6).

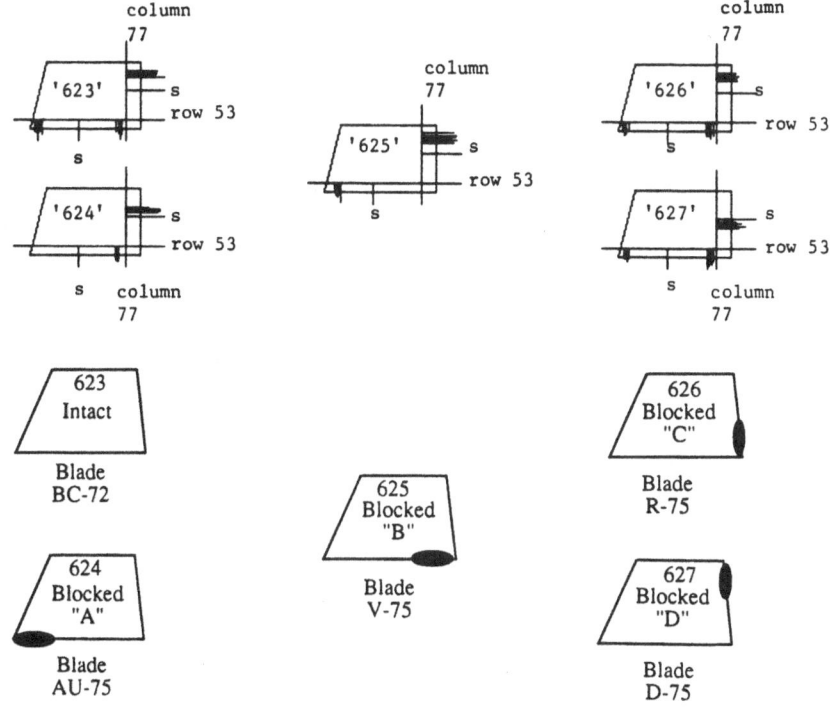

Fig. 5.14. Air rig: automatic interpretation of the inspection procedure (see text).

Following peak detection and location, the peak width (at 3 dB of the maximum) is computed (Fig. 5.13b, c). In Fig. 5.14, sketches of blades for all the categories ("A" to "D") are plotted; analysed rows and columns are also indicated. Also shown on these sketches are the position and the 3 dB width of the peaks detected by the algorithm for all the 78 blades tested. The peak widths are indicated by the length of the line segments while the positions of detected peaks are indicated by the location of these line segments. The line segments marked with an "s" are separator marks used to differentiate the five blade categories. The positions of the separators are computed by statistical analysis (average and standard deviation) for the considered cases. In Fig. 5.14, it is clearly seen that the distinction between intact (experiment 623), blocked "A" (experiment 624), blocked "B" (experiment 625) and blocked "D" (experiment 627) categories is easily made because of the presence or absence of detected peak(s) represented as line segment(s) at a particular location on one side or the other of a given separator. Distinction of blocked "C" (experiment 626) category is also possible by considering the 3 dB peak width along the column of interest (Table 5.1). Using this classification scheme, all the 78 blades were classified correctly.

The choice of the analysed row (here number 53) and column (here number 77) was not found to be critical. An improved SNR can even be obtained by averaging a few (for instance three) rows (or columns) together prior to the analysis. If blades of identical shapes are inspected and clamped in the same position with the blade holder, the analysed row and column stay the same.

Table 5.1.

	3 dB peak width along column 77		
	Average	Standard deviation	Acceptable separator
Blocked "C" category	12.9	1.0	
All other categories	22.2	4.1	15

5.3.3 Discussion

Studies reported in this section show how internal thermal thermography can be implemented for the detection of blocked channels in turbine blades. The same kind of analysis can be implemented for other hollow structures. It was shown how a computerized water or air rig associated with dedicated image processing could sort defective blades automatically. Investigations done on the suction side of the blades revealed similar thermal signatures to those depicted in the figures of this chapter and which correspond to experiments recorded on the pressure side.

Although both stimulation media (air and water) permits us to achieve similar results, air is probably a more advantageous choice since it avoids contamination of the inspected part with a foreign substance (water). On the other hand, as stated previously, due to its good thermal capacity, water analysis yields improved thermal contrast. If air analysis is selected, greater flow and higher temperature differential (for cool– warm thermal stimulation) is needed to obtain acceptable thermal contrasts with respect to the same procedure undergone with a water rig.

Conclusion. In this chapter, we have studied the internal thermal stimulation approach for the inspection of hollow workpieces. This approach provides enhanced thermal contrast due to the high temperature differential between the hot and cold stimulating flow (liquid or gas) inside the inspected structure. Inspection proceeds during the thermal transient (warm to cold or the reverse). Since the thermal perturbing device is inside the component under inspection, it does not generate any radiative noise such as in the case of a hot external thermal perturbing device (e.g. lamps); this is another advantage of this approach. We have also reviewed how image processing techniques of Chap. 4 could be applied for the enhancement and interpretation of recorded thermograms.

Chapter 6

Quantitative Analysis of Delaminations

6.1 General Considerations

In the preceding chapters, we studied infrared thermography, its related theory as well as methods and image processing techniques necessary to investigate materials and structures. We also concentrated on defect detection. Once defects have been located, it is interesting to characterize them quantitatively in order to judge their severity. This sixth chapter is dedicated to this study.

In fact this is known as the *inverse problem*. From the measured thermal, temporal and spatial response of a defect detected by TNDE using methods described in the preceding chapters, we want to evaluate defect size, depth and thermal resistance. In Sect. 2.2, the *direct problem* was studied by obtaining, through thermal modelling, the thermal response for a defect of known geometry.

In this chapter, we will first present, in Sect. 6.2, some accepted methodologies to solve the inverse problem; in Sect. 6.3, we will discuss image processing techniques and experimental procedures needed for the quantitative characterization.

6.2 The Inverse Problem

6.2.1 A Practical Numerical Approach

Many research groups work on inverse problem solving using thermographic information, that is from an experimental TNDE data set, they extract quantitatively subsurface defect properties such as depth, thermal resistivity and size. This is an important issue in order to evaluate quantitatively the severity of a damaged area. Many papers have been published on this issue. Although we do not pretend to present here an exhaustive list of the research groups active in this field, the following list gives particular examples: Williams et al. (1980); Cielo (1984); Sayers (1984); Hsieh and Kassab (1986); Balageas et al. (1987c); Kassab and Hsieh (1987); Degiovanni (1988); Houlbert (1991); Krapez (1991); Krapez and Cielo (1991); Rantala and Hartikainen

(1991); Beaudoin et al. (1985); Beaudoin and Bissieux (1992); Vavilov (1992). Recently, authors have started to rely on neural networks for inversion of NDE data (e.g. Mann et al. 1991). In order to study this inverse problem issue we decided, in the context of this book, to select a practical numerical approach which can be programmed in conjunction with the image processing techniques that will be reviewed in Sect. 6.3. Interested readers are invited to refer to the authors cited above for the complete mathematical analysis.

As a first approach to this matter we will go back to Sect. 2.2.2 where a heat transfer model was discussed. We recall that this model is for a specimen with a subsurface defect; the modelled geometry for an inclusion-type geometry (thermal resistance-type defect) in cylindrical coordinates is shown in Fig. 6.1. The subsurface defect is simulated by having a different thermal resistance with respect to the surrounding bulk material. Other geometries are possible, for instance the modelling of back drilled holes considered as subsurface defects when detected from the front surface.

As an application example of the discussed methods, we will now study the thermal behaviour of graphite epoxy specimens having the same geometry as in the model of Sect. 2.2.2, but with different thermal resistance R_{def} and depth z_{def}. We recall that for the example reviewed in Sect. 2.2.2, a graphite epoxy plate was modelled with a defect

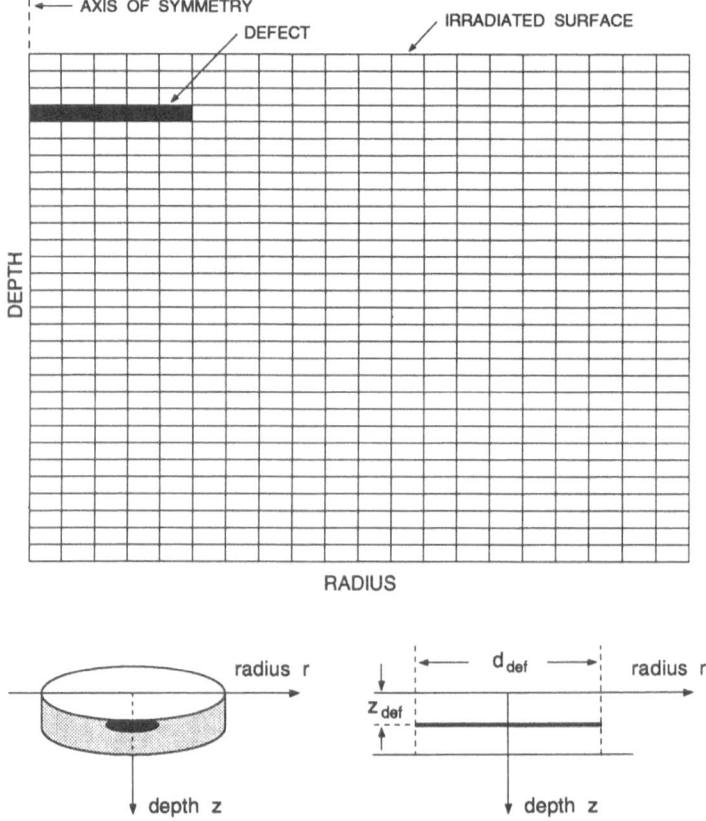

Fig. 6.1. Model for an inclusion-type geometry in cylindrical coordinates; parameters of interest for the defect are also shown.

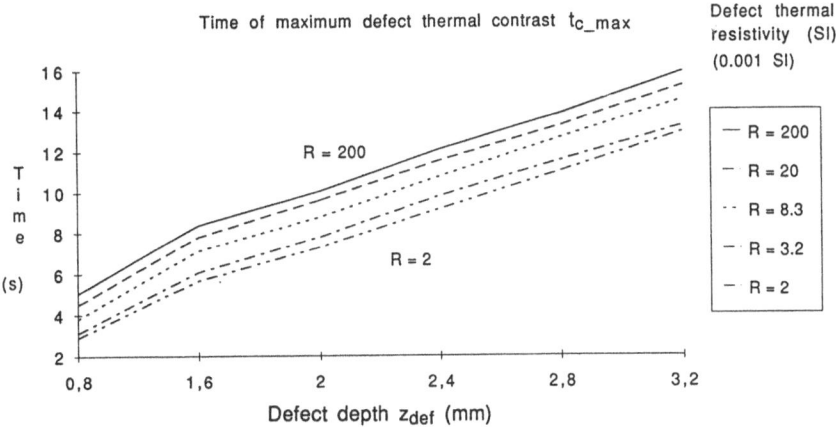

Fig. 6.2. Computed results obtained using model of Sect. 2.2.2 (graphite epoxy specimens); time when thermal contrast reaches a maximum over the defect is plotted for various defect depths and defect thermal resistivity.

thickness corresponding to the thickness of one cell of the mesh. Defect thickness f_{def} can be associated with defect thermal resistance R_{def} by means of the following expression:

$$R_{def} = \frac{f_{def}}{K_{def}} \qquad (m^2 \,°C/W^{-1}, \text{ or SI}) \qquad (6.1)$$

where K_{def} is defect conductivity. To give an idea, if an air layer of 200 μm thickness is modelled, since K_{def} is about $0.024\,W\,m^{-1}\,°C^{-1}$ (Table 1.1), the corresponding thermal resistance will be about 8.3×10^{-3} SI.

For the specimens we will study, two parameters are varied, defect depth z_{def} which varies from 0.8 to 3.2 mm (layer 4 to layer 16 in model of Sect. 2.2.2) while thermal resistance of the defects is changed from $R_{def} = 2 \times 10^{-3}$ to 200×10^{-3} SI. Figures 6.2 and 6.3 summarize computed results obtained for the discussed specimens. In these figures we plot both the time when thermal contrast reaches a maximum over the defect, parameter t_{c_max} (Fig. 6.2) and the value of this contrast (Fig. 6.3), parameter $C_{\_max}$. Since only six depths are considered, the curves, particularly in Fig. 6.3, are not as "smooth" as they should be. Considering the geometry of Fig. 6.1, temperature evolution is observed (Fig. 6.4a) both at defect centre (upper left cell) and far from the defect over a sound area where the effect of the subsurface defect is negligible (upper right cell). This allows us to compute the thermal contrast over the defect, which is given by (Fig. 6.4b)

$$C(t) = T_{def}(t) - T_{sound}(t) \qquad (6.2)$$

Variable t is introduced since thermal contrast varies with time.

On the example of Fig. 6.4b, it is noted that thermal contrast experiences slow variation close to its maximum C_{max}, consequently an uncertainty is associated with the estimation of this parameter and also with the associated time of maximum thermal contrast t_{c_max}. Extraction of the time to reach half the maximum thermal contrast before or after the contrast peak is performed with greater accuracy (Vavilov et al. 1992). These parameters are denoted by $t_{c\_1/2max}$ and $t_{c\_max1/2}$.

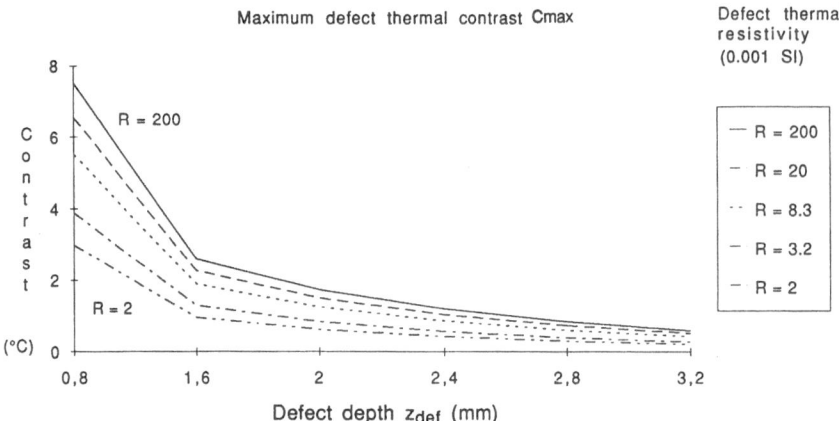

Fig. 6.3. Computed results obtained using model of Sect. 2.2.2 (graphite epoxy specimens); value of the maximum thermal contrast is plotted for various defect depths and defect thermal resistivity.

From data like those used for Fig. 6.2 or 6.3, it is possible to fit functions through standard mathematical packages such as Mathematica™. For instance, using this mathematical package, we computed a function to estimate t_{c_max} with depth z_{def} and thermal resistance R_{def} as input parameters:

$$t_{c\_max} = 1.705 + 1.881 z_{def} + 0.921 z_{def}^2 + 0.000799 z_{def} R_{def}$$
$$+ 0.00138 R_{def} - 8.164 \times 10^{-9} R_{def}^2 \qquad (6.3)$$

Note that for Eq. (6.3), we selected a quadratic function; this is in agreement with Eq. (1.1). In the same fashion, it is also possible to derive functions enabling the inversion process. For instance, combining data both for t_{c_max} (Fig. 6.2) and C_{max} (Fig. 6.3), we can separate the variables R_{def} and z_{def}. The following expression is obtained for z_{def} with C_{max} and t_{c_max} as inputs; this form was originally proposed by Balageas et al. 1987c):

$$z_{def} = 0.6722 \sqrt{t_{c\_max}} (C_{max})^{-0.258} \qquad (6.4)$$

Refer to the information given at the end of this section to help find the parameters of Eq. (6.4).

This kind of analysis, performed by first modelling the process to be analysed and then inverting the computed data through mathematical procedures, is a simple though efficient approach to the inverse problem. In order to be much more efficient, however, it is possible to combine more information available on the detected defect such as $t_{c\_1/2max}$, $t_{c\_max 1/2}$ which are extracted from the temperature evolution curve of the specimen. This extraction process is explained in more detail in Sect. 6.3 below). Of course, this kind of analysis yields better results if more cases are analysed in the direct modelling process. In the present example, only 36 cases were modelled on a restricted span of depths z_{def} and thermal resistances R_{def}.

Equation (6.4) provides accurate figures to a few percent on a restricted span of values. Such a figure is generally the best we can expect from this kind of analysis. This is acceptable, especially if we consider the many other variables present when actual TNDE experiments are performed. For instance, uneven heating, approximate values for thermal properties of the specimen, variability of the thermal losses

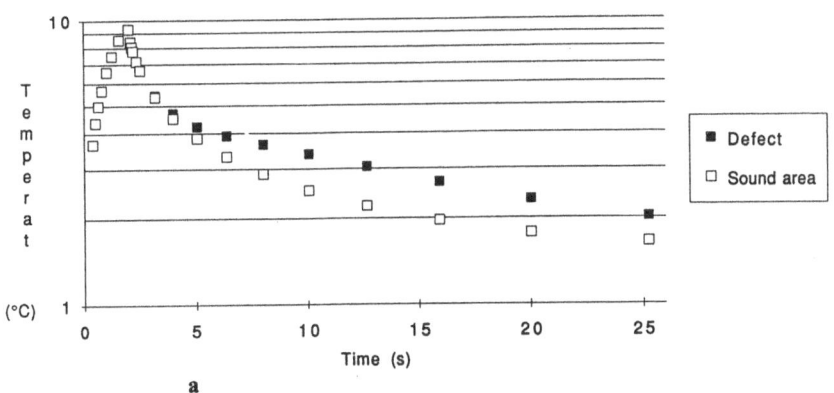

a

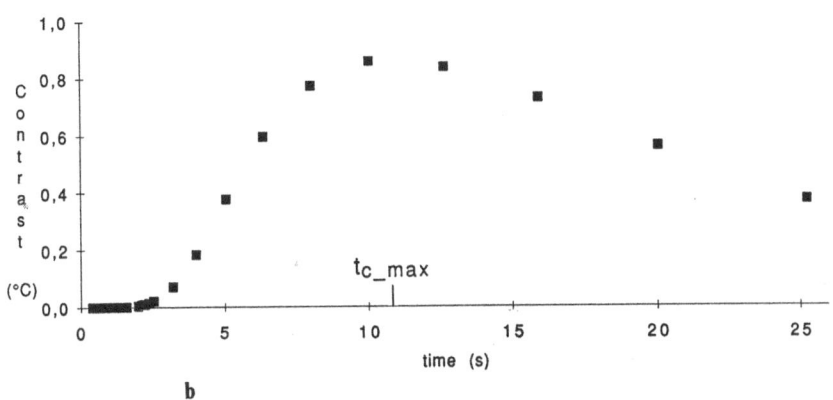

b

Fig. 6.4. Typical example of **a** temperature and **b** contrast evalution curves obtained with model of Sect. 2.2.2 (graphite epoxy specimen, geometry $z_{def} = 2.4\,mm$, $d_{def} = 20\,mm$ and $R_{def} = 8.3 \times 10^{-3}\,SI$). Plot of **b** is obtained using data of **a**. Note: same modelled specimen as for Fig. 2.7.

(convection, radiation), difficulty in estimating the emissivity of the surface under inspection, to mention only a few; also plant floor conditions are often very different from laboratory conditions, which obviously restricts the limit of accuracy such model–experiment comparisons can achieve. This is especially the case when surface-wide inspection is performed. In fact, better agreement is often obtained from point-by-point analysis due to the better controlled thermal stimulation conditions. Although convenient, the use of inversion functions such as Eq. (6.4) may not prove accurate enough in some circumstances. For these situations, it is possible to proceed with inverse interpolation using computed data obtained with the direct model (Krapez et al. 1991).

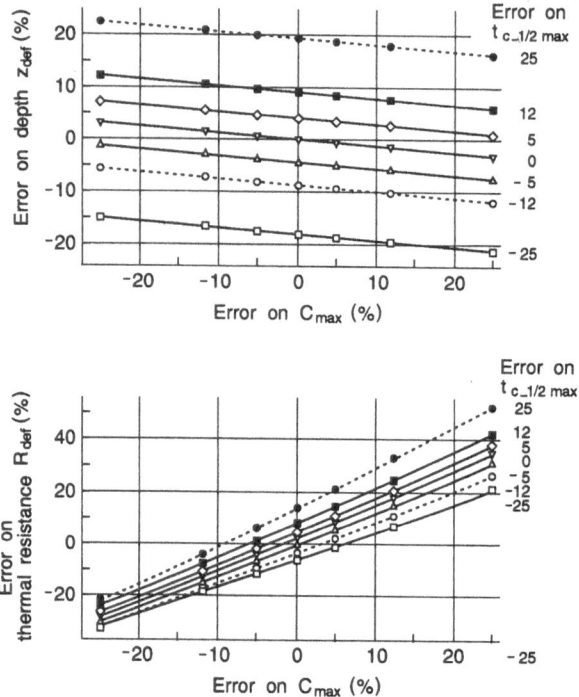

Fig. 6.5. Effect of the uncertainties introduced on input parameters such as noise on experimental start-up data. Cases of defect depth z_{def} and thermal resistance R_{def} are shown (graphite epoxy specimens).

A good validation procedure for the inversion functions consists of feeding back data obtained with the direct model into these functions: this gives information about the best degree of agreement we should expect and also determines the effect of possible uncertainties in the input parameters. Such a performance analysis is an important tool. For instance, Fig. 6.5 presents typical results of such analysis. The curves show that generally effects are linear and proportional. For evaluation of thermal resistance R_{def} it is noted the error tends to amplify, consequently inversion procedures should yield greater uncertainties on the thermal resistance than on the depth or size evaluation.

In order to validate the (direct) model, it is necessary to perform actual experiments under controlled laboratory conditions and compare experimental results with those actually obtained with the model. Ideally the samples tested have to be destroyed after inspection in order to determine exactly the nature and geometry of the detected flaws. If a sufficiently large collection of samples is tested, it is possible to derive a statistical measure of confidence on the inverse functions. This permits quantitative measurements to be obtained with a known degree of uncertainty when actual inspection sessions take place.

In this section we have considered the depth z_{def} of the defect; in the next section we will discuss more about the estimation of both thermal resistance R_{def} and defect size which can be associated with corresponding defect diameter d_{def}.

Note on Eq. (6.4)

Common mathematical packages cannot fit data based on equations of the kind of Eq. (6.4). Using logarithms, such formulas simplify to a line whose parameters can be evaluated using a simple pocket calculator (with the regression function)! Let the formula be expressed by

$$z_{def} = A\sqrt{t_{c\_max}}(C_{max})^n \tag{6.5}$$

Taking the logarithm on both sides and rearranging the terms:

$$\log(z_{def}) - \log(\sqrt{t_{c\_max}}) = \log(A) + n\log(C_{max}) \tag{6.6}$$

$$\log\left(\frac{z_{def}}{\sqrt{t_{c\_max}}}\right) = \log(A) + n\log(C_{max}) \tag{6.7}$$

$$\log\left(\frac{z_{def}}{\sqrt{t_{c\_max}}}\right) = B + n\log(C_{max}) \tag{6.8}$$

If, in the fitting procedure, data are expressed as $\log(z_{def}/\sqrt{t_{c\_max}})$ and $\log(C_{max})$ which at this stage are known, it is only necessary to compute B and n; Eq. (6.8) corresponds to the equation of a line and we finally obtain

$$z_{def} = \log(e^B)\sqrt{t_{c\_max}}(C_{max})^n \tag{6.9}$$

Obviously the constant $\log(e^B)$ is easily calculated; Eq. (6.9) is of the same form as Eq. (6.5).

6.2.2 Normalized Variables

Use of normalized variables allows us to generalize the numerical approach presented in the previous section and apply these concepts to a wider span of subsurface defect geometries ($z_{def}, R_{def}, d_{def}$).

In fact, investigations and results presented in the previous section are only valid for a very specific geometry. In some instances, it is interesting to rely on broader geometries. Such studies make use of normalized parameters, the Fourier numbers Fo and the Biot numbers Bi.

Characterization of Defect Depth and Thermal Resistance

First, let us introduce the effusivity $e(t)$ which describes the temperature evolution of the sample surface; for times t such as $\alpha t/L^2 < 0.1$, it can be expressed by (Balageas et al. 1987c, Eq. 2).

$$e(t) = \frac{Q}{\Delta T(t)\sqrt{\pi}\sqrt{t}} \tag{6.10}$$

where Q is the density of the pulsed energy, ΔT is the temperature evolution, L is the material thickness and α is the material thermal diffusivity, $\alpha = K/\rho C$, ρ is the density, C is the specific heat and K is the thermal conductivity (cf. Table 1.1).

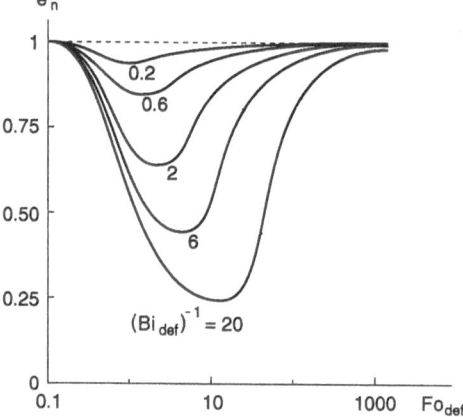

Fig. 6.6. Evolution of normalized effusivity e_n over the defect (specimen of graphite epoxy made of two layers with identical thermal properties; the defect, a delamination, is characterized by a specific Biot number Bi_{def}.

The Fourier number Fo_{def} for a defect depth z_{def} is expressed as

$$Fo_{def} = \frac{\alpha t}{z_{def}^2} \tag{6.11}$$

and the Biot number Bi_{def} as

$$Bi_{def} = \frac{z_{def}}{KR_{def}} \tag{6.12}$$

In Fig. 6.6, we show the evolution of normalized effusivity e_n over the defect for some values of Bi_{def}. The normalized effusivity e_n is given by

$$e_n = \frac{e}{e_{material}} = \frac{e}{\sqrt{K\rho C}} \tag{6.13}$$

It is interesting to note that considering a given combination of defect properties (R_{def}, z_{def}), the effusivity curve is completely characterized, for instance by its minimum $(Fo_{def,min}; e_{n,min})$, by its half-decrease point $(Fo_{def,1/2min}; e_{n,1/2min})$ or by its half-rise point $(Fo_{def,min1/2}; e_{n,min1/2})$. Note that this is similar to the analysis of Sect. 6.2.1 when we considered the time of maximum or half contrast rise.

From these parameters, Balageas et al. (1987c) established interesting inverse functions in the case of subsurface delaminations detected in layered materials:

$$e_{n,min} = [Fo_{min}]^{-0.528} \tag{6.14}$$

$$z_{def} = \sqrt{\alpha}\sqrt{t_{min}}[e_{n,min}]^{0.95} \tag{6.15}$$

where t_{min} is the time where the effusivity curve is minimum. This last relationship is similar to Eq. (6.4) studied in the previous section. Another useful inverse function is

$$z_{def} = 1.61\sqrt{\alpha}\sqrt{t_{1/2min}}[e_{n,1/2min}]^{0.85} \tag{6.16}$$

For cases where the determination of the minimum of the effusivity curve is not possible with sufficient precision, the use of half-decrease point is preferred.

Thermal resistance characterization is also possible using a similar approach (Delpech and Balageas 1991). However, it is not possible to express R_{def} in a form as compact as for the defect depth z_{def}. Balageas et al. (1987c) first introduce the function ψ relating $e_{n,min}$ and Bi_{def}^{-1}, (Fig. 6.6):

$$e_{n,min} = \psi(Bi_{def}^{-1}) \tag{6.17}$$

and then they found:

$$R_{def} = \frac{1}{e_{material}} \sqrt{t_{min}} [e_{n,min}]^{0.95} \psi^{-1}(e_{n,min}) \tag{6.18}$$

or using the half decrease point:

$$R_{def} = \frac{1.61}{e_{material}} \sqrt{t_{1/2min}} [e_{n,1/2min}]^{0.95} \psi^{-1}(e_{n,min}) \tag{6.19}$$

Notice the similarity with Eqs (6.15) and (6.16). Using similar numerical methods to those presented in Sect. 6.2.1, inverse functions can be derived for particular materials and geometries.

Characterization of Defect Size

It is possible to address the characterization of defect size (corrsponding to an equivalent defect diameter d_{def}) in a similar fashion to defect depth z_{def} and defect thermal resistance R_{def}. Krapez and Cielo (1991) found that position of the steepest temperature gradient recorded on the sample surface either at time of maximum contrast t_{c_max} or at half rise time $t_{c\_1/2max}$ is very close to the position of the subsurface defect border. Measurement of this gradient is called the apparent defect diameter d_{app}. The normalized diameter d_n is then defined by $d_n = (d_{def}/z_{def})(\alpha_z/\alpha_r)^{1/2}$ where α_z and α_r are the thermal diffusivities in the transverse and radial direction of the inspected specimen respectively.

Heat transfer modelling allows us to construct curves like those shown in Fig. 6.7 relating d_{app} and d_n for various normalized defect resistances $R_n (R_n = R_{def} K/z_{def})$, where K is the thermal conductivity and z_{def} is the defect depth. From such curves,

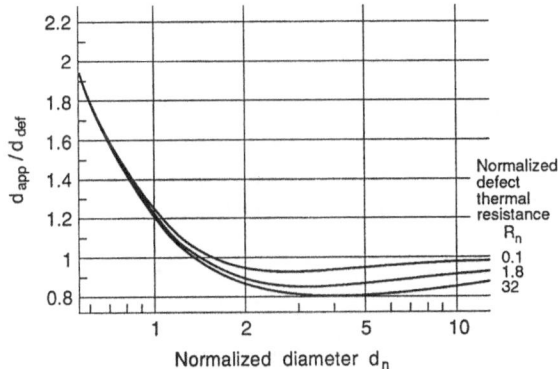

Fig. 6.7. Computational results showing the relation of the d_{app} and d_n for various normalized defect resistances R_n, graphite epoxy specimens.

recovery of the value of d_{def} once values of z_{def}, R_{def} and d_{app} (recorded either at t_{c_max} or at $t_{c\_1/2\,max}$) have been established is possible using similar techniques as described in the previous sections; inverse functions $d_{def} = g(z_{def}, R_{def}, d_{app})$ can be derived. Of course d_{app} depends upon both z_{def} and R_{def}: a large deep defect can have the same d_{app} figure as a shallow smaller subsurface flaw (this is clearly seen in Fig. 4.29j (colour section); for instance in the case of the 10 mm diameter defects on image "P_D.3", different depths of 1.12 mm (top row) and 2.25 mm (bottom row) lead to different d_{app} values).

If the subsurface defect is not circular in shape, the corrective factor d_{app}/d_{def} may be applied to the apparent defect shape (steeper temperature gradient on the sample surface) to recover the correct flaw shape.

In the next section, we will study the various procedures and methods necessary to the extraction of experimental parameters.

6.3 Experimental Procedure: Thermogram Processing

In the last section, we described practical methods useful to solve the inverse problem; we came to the conclusion that knowledge of a few parameters extracted along the temperature evolution curve, the curve recorded at the centre of the detected subsurface flaw on the specimen surface, enables evaluation of the defect characteristics, mainly the depth z_{def}, the thermal resistance R_{def}, and the size corresponding to an equivalent defect diameter d_{def} (Fig. 6.1). The parameters of interest are: the value of the maximum thermal contrast C_{max}, time of maximum contrast t_{c_max}, time of half-rise contrast $t_{c\_1/2max}$ and time of half-decay $t_{c\_max1/2}$. Figure 6.8 shows these parameters over an experimental temperature evolution curve.

A special experimental procedure is required to extract, from the recorded thermogram sequence, parameters $t_{c\_1/2max}$, t_{c_max}, $t_{c\_max1/2}$, and also to evaluate subsurface flaw size. We review now the required experimental procedure for such quantitative thermogram processing.

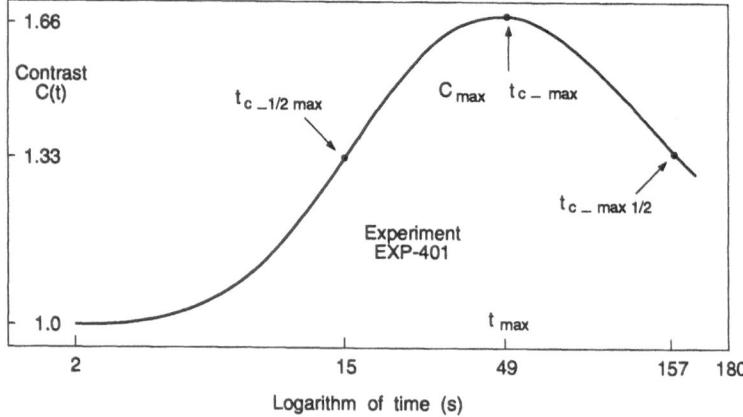

Fig. 6.8. Experimental thermal contrast evolution curve; parameters of interest are indicated (Plexiglass™ plate with back hole 10 mm diameter, 1.62 mm below the front surface, see Sect. 6.4 for more details).

6.3.1 Thermal Contrast

Let us consider a surface having an emissivity $\varepsilon \sim 1$ initially at temperature T_0. The ambient environment (env) is at temperature T_a. Suppose we heat this surface so that a temperature T_{def} is observed over a defect and a temperature T_{soa} is obtained over a sound area. The radiance signal picked up by the thermal imager and available in image format is given by, from Eqs (2.6):

$$I_{img\_def}(T_{def}) = \varepsilon I_{def}(T_{def}) + (1-\varepsilon)I_{env}(T_a) \tag{6.20}$$

$$I_{img\_soa}(T_{soa}) = \varepsilon I_{soa}(T_{soa}) + (1-\varepsilon)I_{env}(T_a) \tag{6.21}$$

$$\underbrace{\hspace{3cm}}_{\substack{\text{Contribution} \\ \text{from the surface}}} \quad \underbrace{\hspace{3cm}}_{\substack{\text{Thermal reflections} \\ \text{from the surroundings}}}$$

Before heating (at $t = 0$) the surface was at uniform temperature T_0; the signals were

$$I_{img\_def}(T_0) = \varepsilon I_{def}(T_0) + (1-\varepsilon)I_{env}(T_a) \tag{6.22}$$

$$I_{img\_soa}(T_0) = \varepsilon I_{soa}(T_0) + (1-\varepsilon)I_{env}(T_a) \tag{6.23}$$

If we subtract Eqs (6.20)–(6.22) and (6.21)–(6.23) to suppress the adverse contributions from the surrounding environment:

$$I_{img\_def}(T_{def}) - I_{img\_def}(T_0) = \varepsilon[I_{def}(T_{def}) - I_{def}(T_0)] \tag{6.24}$$

$$I_{img\_soa}(T_{soa}) - I_{img\_soa}(T_0) = \varepsilon[I_{soa}(T_{soa}) - I_{soa}(T_0)] \tag{6.25}$$

we take the ratio of Eqs (6.24)/(6.25):

$$C'' = \frac{I_{def}(T_{def}) - I_{def}(T_0)}{I_{soa}(T_{soa}) - I_{soa}(T_0)} \tag{6.26}$$

The four terms in Eq. (6.26) can be converted to temperature using Eqs (2.11), (3.19) and (3.20) to obtain the contrast C':

$$C' = \frac{T_{def} - T_{0,def}}{T_{soa} - T_{0,soa}} = \frac{\Delta T_{def}}{\Delta T_{soa}} \tag{6.27}$$

We can now add both the spatial variations (since the thermal contrast may change for every pixel (i, j) in the image) and the time variations t (since the thermal contrast evolves during the experiment) to obtain the thermal contrast equation:

$$C(i, j, t) = \frac{\Delta T_{def}(i, j, t)}{\Delta T_{soa}(t)} = \frac{T_{def}(i, j, t) - T_{def}(i, j, t=0)}{T_{soa}(t) - T_{soa}(t=0)} \tag{6.28}$$

We recall that $T_{soa}(t)$ is the temperature averaged over the whole sound area (soa). Equation (6.28) gives the thermal contrast, which is the temperature differential normalized by the temperature differential over a nondefect area. A unit value will be obtained over an area without defect if the disturbances from emissivity and thermal reflections stay constant during the experiment. Equation (6.28) combines together both the temporal and the spatial reference methods reviewed in Chap. 5 and is a standard definition of thermal contrast (Lau et al. 1990).

6.3.2 Logarithmic Time Scale

Equation (6.28) allows us to evaluate the thermal contrast above the detected subsurface defect for any time t. In Chap. 3, we discussed practical limitations of image acquisition systems, both on the rate of acquisition and on the total amount of memory available to record a complete thermal experiment. Taking into account these limitations, it is necessary to rely on a *temporal scale* in order to perform inspections which can last up to 15 min for some low thermal conductivity samples.

A logarithmic time scale is selected for two reasons. First, with temporal intervals getting larger with time, we can proceed to inspection sessions which span over long periods of time while having, at the beginning, short time intervals. This allows us to catch fast thermal events which are likely to happen at the beginning of the thermal evolution curve. In other words, the use of a logarithmic time scale permits us to make linear the quadratic law of thermal diffusion (Eq. (1.1)).

Second, as time passes, thermal contrasts weaken because of the three-dimensional diffusion of the thermal front which tends to make uniform the surface temperature distribution, thus increasing the perturbing effect of the noise. Large time intervals containing a large number of images which are averaged together allow the SNR to be increased without distorting the thermal contrast evolution curve since, for these large time values, thermal events change slowly.

Disjoint Case

The logarithmic time scale is divided into N zones from time t_0 after starting the thermal perturbation (at $t = 0$) to t_f at the end of the experiment:

$$
\begin{array}{ccccccc}
i=0 & i=1 & \cdots & i & \cdots & i=N-1 \\
\Big| \quad \Big| & \Big| \quad \Big| & & \Big| \quad \Big| & & \Big| \quad \Big| \\
t_{0-} \quad t_0 & t_{1-} \quad t_1 & \cdots & t_{i-} \quad t_i & \cdots & t_{f-} \quad t_f
\end{array}
$$

Two others parameters are system dependent: q is the image acquisition rate, this is the time required to frame-grab and temporarily save the image in fast memory (images are later saved on slower high volume memory media such as a hard disks or diskettes) and NM is the maximum number of images the fast temporary memory can hold.

The logarithmic time interval is given by

$$
\Delta \log = \frac{\log t_f - \log t_0}{N-1} = \frac{1}{N-1} \log\left(\frac{t_f}{t_0}\right) \tag{6.29}
$$

and for zone t_i corresponding to zone i of the temporal scale, we have:

$$
\log t_i = \log t_0 + i\Delta \log \qquad i = 0, 1, \ldots, N-1
$$
$$
t_i = t_0 \left(\frac{t_f}{t_0}\right)^{i/N-1} = t_0 \mu^i \quad \text{with} \quad \mu = \left(\frac{t_f}{t_0}\right)^{1/N-1} \tag{6.30}
$$

Low borders of time zones are given by $t_{0-}, t_{1-}, \ldots, t_{i-}, \ldots, t_{n-}$ so that zone i spans from t_{i-} to t_i. Let us introduce λ which specifies zone width:

$$
\log t_{i-} = \log t_i - \log \lambda
$$
$$
t_{i-} = \frac{t_i}{\lambda} \tag{6.31}
$$

The parameter λ is evaluated by an iterating process to ensure the maximum possible number of images NM will be recorded knowing that we cannot grab a *fraction* of an image. Obviously, there is also at least one image per zone. In the following formula "1" is added since as soon as the image acquisition process is started, it must be ended:

$$NM \approx NM_{estimated} = \left\{ \frac{t_0 - t_{0-} + 1}{q} + \cdots + \frac{t_i - t_{i-} + 1}{q} + \cdots + \frac{t_f - t_{f-} + 1}{q} \right\}$$

which simplifies using Eqs (6.30) and (6.31):

$$NM \approx NM_{estimated} = \frac{t_0}{q\left(1 - \frac{1}{\lambda_{first\ run}}\right)} \left\{ \left(\frac{1 - \mu N}{1 - \mu}\right) + N \right\} \qquad (6.32)$$

By isolating λ in Eq. (6.32), a starting value $\lambda_{first\ run}$ can be computed. Secondly, the exact value of λ is obtained by adjusting its value and computing the *exact* number of images until NM images are obtained, $NM_{estimated} \rightarrow NM$:

$$NM_{estimated} = \text{Integer Part}\left[\left(\frac{t_0}{q}\right)\left(1 - \frac{1}{\lambda}\right) + 1 \right] + \cdots$$

$$+ \text{Integer Part}\left[\left(\frac{t_f}{q}\right)\left(1 - \frac{1}{\lambda}\right) + 1 \right] \qquad (6.33)$$

There is another constraint on λ since time zones cannot overlap, so we must have

$$t_0 < t_{1-} \qquad (6.34)$$

thus with Eqs (6.30) and (6.31):

$$\lambda < \mu \qquad (6.35)$$

In Table 6.1 we show an example of logarithmic time scale computed using Eqs (6.30) to (6.33); time zones are disjoint.

It is not always possible to respect constraints on all parameters (t_0, t_f, N, NM and q) simultaneously. This means acquisitions are done on a non-stop basis, time zones become contiguous ($\lambda = \mu$) and the algebraic derivation of the time zones is slightly different, as we will see now.

Continuous Case

In the continuous case the end of a zone equates with the beginning of the next; using Eqs (6.30 and 6.31) we have

$$t_{i-1} = t_{i-} \qquad (6.36)$$

$$t_0 \mu^{i-1} = \frac{t_0 \mu^i}{\lambda} \qquad (6.37)$$

$$\frac{t_0 \mu^i}{\mu} = \frac{t_0 \mu^i}{\lambda} \qquad (6.38)$$

$$\mu = \lambda \qquad (6.39)$$

Before proceeding to the actual time zone computation, it is necessary to check if the parameters supplied by the user (t_0, t_f, N) allow us to obtain at least one image in each zone. For this to occur, following the acquisition of one image we immediately proceed with the next zone. However, the starting time of the next time zone must

Table 6.1. Illustration of disjoint time zone case

Input parameters		
NM (maximum number of images)		284
q (acquisition + temporary storage, one image)		0.27 s
N (number of zones)		10
t_0 (start of acquisition)		1 s
t_f (end of experiment)		100 s
Computed parameters		
μ		1.668101
λ (first run)		1.422476
$NM_{estimated}$ (first run)		279
λ (42nd run)		1.432348
$NM_{estimated}$ (42nd run)		284

Zone	$t-$ (s)	t (s)
0	0.70	1.00
1	1.16	1.67
2	1.94	2.78
3	3.24	4.64
4	5.41	7.74
5	9.02	12.92
6	15.04	21.54
7	25.09	35.94
8	41.85	59.95
9	69.82	100.00

not be already exceeded after the last image acquisition of the preceding time zone:

$$\frac{t_0}{\lambda} \quad t_0 \qquad \frac{t_1}{\lambda} \quad t_1 \quad \cdots$$

From this we immediately see that

$$
\begin{array}{ccccc}
\text{Start of} & + & \text{Time to} & < & \text{Beginning} \\
\text{acquisition} & & \text{grab} & & \text{of next zone} \\
\dfrac{t_0}{\lambda} & + & q & < & \dfrac{t_1}{\lambda}
\end{array}
\tag{6.40}
$$

with Eqs (6.30), (6.31) and (6.39):

$$t_0 < (t_0 - q)\mu \tag{6.41}$$

and finally

$$\frac{t_0}{t_0 - q} < \mu \tag{6.42}$$

If Eq. (6.42) is respected this means there is no overlapping zone and the computation can be done as previously using Eqs (6.30) to (6.33) and Eq. (6.35) will be respected. If Eq. (6.42) is not respected the contraints imposed on the parameters are too strong for the given value of q (q is the time to frame-grab an image; it is installation dependent). As mentioned before, this corresponds to the nonstop acquisition case and t_f and N must be adjusted accordingly. In this case, since we want at least one image per zone we are at the limit of Eq. (6.42) and parameter μ is computed as follows:

$$\mu = \frac{t_0}{t_0 - q} \qquad (6.43)$$

The time for the end of experiment t_f is changed:

$$t_f = \text{minimum} \ [t_f \ (\text{specified by user}), \ t_0 + NM \ q] \qquad (6.44)$$

and the number of zones is computed to satisfy Eq. (6.43) with (6.30):

$$N = \text{Rounded} \left\{ 1 + \frac{\log\left(\dfrac{t_f}{t_0}\right)}{\log \mu} \right\} \qquad (6.45)$$

In this case, all the zones will be contiguous. In Table 6.2 we show an example of logarithmic time scale computed using Eqs (6.30), (6.31), (6.43), (6.44) and (6.45); time zones are contiguous. Notice that in Table 6.2, the last zone ends at $t = 81.94$ s instead of 77.68 s because in Eq. (6.45), we must have a whole number of zones.

We may now summarize time zone computation once the user has specified the input parameters t_0, t_f, N, NM and q.

Table 6.2. Illustration of contiguous time zone case

Installation parameters	
NM (maximum number of images)	284
q (acquisition + temporary storage, one image)	0.27 s
Desired parameters	
N (number of zones)	30
t_0 (starting of acquisition)	1 s
t_f (end of experiment)	100 s
Computed parameters:	
$\mu = \lambda$:	1.369863
t_f (end of experiment)	77.68 s
N (number of zones)	15

Zone	$t-$ (s)	t (s)
0	0.73	1.00
1	1.00	1.37
2	1.37	1.88
3	1.88	2.57
4	2.57	3.52
5	3.52	4.82
6	4.82	6.61
7	6.61	9.05
8	9.05	12.40
9	12.40	16.99
10	16.99	23.27
11	23.27	31.87
12	31.87	43.66
13	43.66	59.81
14	59.81	81.94

First Eq. (6.42) is checked using Eq. (6.30):

If respected (Disjoint case) use   Eqs (6.30)–(6.33)

Otherwise (Contiguous case) use   Eqs (6.30), (6.31), (6.43), (6.44) and (6.45)

As we saw in Chap. 3, the stimulating heat pulse has a finite duration. It can be demonstrated that in these conditions, the best origin for the time scale is the one corresponding to the barycentre of the heating pulse (Fort et al. 1988; Krapez 1991). The origin of the time scale is thus the barycentre of the heating pulse defined as $t = 0$.

Once the thermal inspection is completed, we obtain a certain number of images in each temporal zone. In order to limit noise, images are averaged in each zone; the idea is to obtain only one resulting image associated with each zone i corresponding to time t_i. This value t_i depends on the acquisition time of all images averaged together in zone i. Since on the temperature history curve (Fig. 1.2), temperature T follows time as $1/\sqrt{t}$ after application of the thermal perturbation (heat pulse), we have

$$t_i = \frac{1}{\left(\dfrac{1}{n_i}\displaystyle\sum_{j=1}^{n_i}\dfrac{1}{\sqrt{t_j}}\right)^2} \qquad (6.46)$$

where n_i is the number of images in time zone i and t_j is the time corresponding to image number j in zone i. For instance if there are three images recorded at time 8, 9, 10 s in the considered zone, the time associated with this zone will be

$$t = \frac{1}{\dfrac{1}{3}\left[\dfrac{1}{\sqrt{8}} + \dfrac{1}{\sqrt{9}} + \dfrac{1}{\sqrt{10}}\right]^2} = 8.94\,\text{s}$$

6.3.3 Practical Computation of $C(t)$, C_{max}, $t_{c\_1/2max}$, t_{c_max}, $t_{c\_max1/2}$

We have now all the elements needed for the experimental computation of the thermal contrast. Once the experiment has been performed based on the techniques reviewed in preceding chapters, if a subsurface flaw is present, using the automatic segmentation algorithm of Sect. 4.4, it is located from the series of recorded thermal images at the position (i_d, j_d) corresponding to the centre of the defect. Thermal contrast $C(t)$ is now computed using Eq. (6.28).

1. Preliminary computations:
   a series of say 15 images is recorded and averaged prior to the beginning of heating in order to obtain the so-called *cold image* (Img0). This image is also converted in temperature following techniques of Sect. 3.3 to obtain $T_0(i, j)$, $\forall i, j$.
   A region far from the defect called r_far comprising s elements is defined in the image field. Also, from cold image Img0 we compute temperature for all elements 1 of this zone to obtain row vector $r\_farT0(1)$, $\forall 1 = 0, 1, \ldots, s - 1$.
   A row vector called RowT0 is computed from the cold image Img0. It is the average, pixel by pixel, of three rows converted into temperature and passing through defect centre; these are rows i_{d-1}, i_d, i_{d+1}:
   For $\forall j, j = 0, 1, \ldots, \text{Maxcol-1}$:

$$\text{RowT0}(j) = \frac{1}{3}\sum_{i=i_d-1}^{i=i_d+1} T0(i, j)$$

2. For every time zone k, thermal contrast $C_{row}(k, j)$ along elements j of row i_d passing by the hottest point of the defect (i.e. defect centre) is computed using Eq. (6.28). In fact this contrast computation is performed on the three rows and the three contrast values are averaged together for each position of the row to obtain the row vector $C_{row}(k, j)$. For $\forall j, j = 0, 1, \ldots$, Maxcol-1, we compute:

$$C_{row}(k, j) = \frac{1}{N_k} \sum_{m=0}^{N_k - 1} \left[\frac{RowT(m, j) - RowT0(j)}{\frac{1}{s} \sum_{l=0}^{l=s-1} [r\_far\,T(m, l) - r\_far\,T0(l)]} \right] \tag{6.47}$$

where $RowT(m, j)$ is the jth element ($j = 0, 1$, Maxcol-1) obtained from the computation of the average of the rows i_{d-1}, i_d, i_{d+1} of the mth image belonging to time zone k converted into temperature $T(m, i, j) \forall i, j$. For $\forall j, j = 0, 1, \ldots$, Maxcol-1, we compute

$$RowT(m, j) = \frac{1}{3} \sum_{i = i_d - 1}^{i = i_d + 1} T(m, i, j)$$

where $r\_far\,T(m, l)$ is the lth element ($l = 0, 1, 2, \ldots, s-1$) of the sound region defined far from the defect and computed in temperature for the mth image in time zone k, and m is the image index for time zone k, $m = 0, 1, 2, N_k - 1$. There are N_k images in time zone k.

The denominator of Eq. (6.47) is a scalar number which divides each element of the numerator. All the temperature computations are done following the procedure described in Sect. 3.3 (Eqs (3.19) and (3.20)). Row vector $C_{row}(k, j), \forall j$ corresponds to thermal contrast of the image row passing through the defect centre for time zone k; time zone k corresponds to an instant given by Eq. (6.46).

Contrast computations thus proceed on an image row basis. From a programming point of view, this format is advantageous since we obtain, for all time zones, an image of the temporal evolution of the thermal contrast where every row (one row per time zone) is given by Eq. (6.47). Since this corresponds to the same format as a standard image [number of rows] × [number of columns], Maxrow × Maxcol (missing rows are set to 0 if the number of time zones is smaller than Maxrow), all display routines, save programs, etc. are the same as for "regular" thermograms. Figure 6.9 shows a typical result: this is the temporal thermal contrast evolution recorded over a drilled hole (5 mm diam., 2 mm beneath the surface in a graphite epoxy plate). There are 40 time zones, from $t_0 = 5$ s to $t_f = 120$ s. In this figure, the rise and decay of the thermal contrast is well evident.

From this plot of the temporal evolution of the thermal contrast ($C_{row}(k, j), \forall j, \forall k$) which is smoothed using techniques described in Sect. 3.2.2, we extract the value of the maximum contrast above the defect:

$$C_{row}(k_{max}, j_{max}) > C_{row}(k, j) \forall k, j \tag{6.48}$$

Where k represents the considered time zone, and j represents the position along the row, generally $j_{max} = j_d$.

The profile of the maximum temporal thermal contrast is then extracted from the plot of the thermal contrast over the defect (Fig. 6.10):

$$C_{row}(k, j_{max}), \quad \forall \text{ time zones } k = 0, 1, (N_k - 1) \tag{6.49}$$

Figure 6.10 shows the profile $C_{row}(k, j_{max})$ obtained in the case of Fig. 6.9. From the plot of Eq. (6.49), $t_{c\_1/2max}, t_{c\_max}, t_{c\_max1/2}, C_{max}$ are extracted. We recall that these

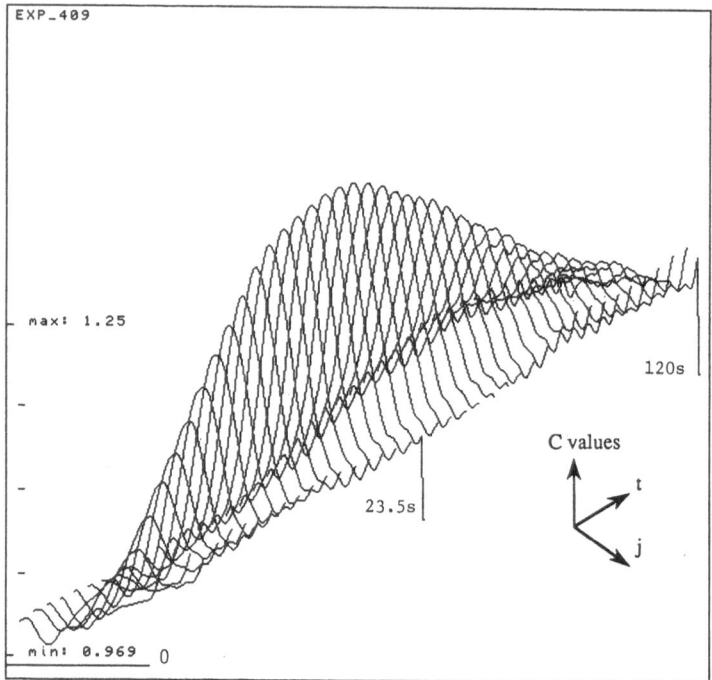

Fig. 6.9. Temporal contrast evolution curve above graphite epoxy plate with defect (hole of 5 mm diam. drilled 2 mm under the front surface). The 40 time zones are shown from 5 to 120 s. Every line plotted corresponds to the thermal contrast along the image rows (average of three) passing just above the defect for a given time zone.

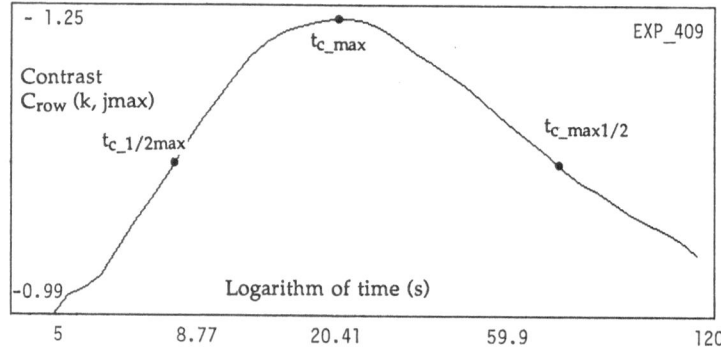

Fig. 6.10. Evolution of the maximum contrast $C_{\text{row}}(k, j\text{max})$ extracted from Fig. 6.9 (see text).

parameters are needed for the inversion procedure of Sect. 6.2. For accurate extraction, it is necessary to extrapolate (a line or a parabola) since we only have values at instants corresponding to time zones k:

C_{max} Maximum contrast

t_{c_max} Instant of maximum contrast. C_{max} and t_{c_max} are obtained by computing the parabola passing through the three or five points having the largest value in the contrast vector of Eq. (6.49), see App. C for details on the computation.

$t_{c\_1/2max}$ Instant at which thermal contrast reaches half its maximum value C_{max}. It is obtained by searching through the contrast vector of Eq. (6.49) for the closest half-rise contrast values and by interpolating on two or four neighbours.

$t_{c\_max1/2}$ Instant at which thermal contrast decays to half its maximum value C_{max}. It is obtained by searching through contrast vector of Eq. (6.49) for closest half-decay contrast values and interpolating on two or four neighbours.

If we take the example of Fig. 6.10, the following values are obtained:

Maximum contrast C_{max}:

$$[5 \text{ points}] = 1.247 \qquad t_{c\_max} = 20.32\,\text{s}$$
$$[3 \text{ points}] = 1.247 \qquad t_{c\_max} = 20.42\,\text{s}$$

Half-rise thermal contrast $t_{c\_1/2max}$:

$$[4 \text{ points}] = 8.744\,\text{s (half contrast} = 1.124)$$
$$[2 \text{ points}] = 8.772\,\text{s}$$

Half-decay thermal contrast $t_{c\_max1/2}$:

$$[4 \text{ points}] = 59.94\,\text{s}$$
$$[2 \text{ points}] = 59.95\,\text{s}$$

From our definition of thermal contrast Eq. (6.28), a value of one corresponds to an absence of defect, that is a state of uniform temperature. For example, in the case of Fig. 6.10, the thermal contrast above the defect decays to 1.05 (i.e. 5% contrast) 2 min after heating because of the temperature which tends to become uniform due to the cooling down of the graphite epoxy plate (i.e. heat spreading diffusion process within the material and surface losses).

Defect shape (d_{app}) is obtained from thermal contrast image C_{img} at the time thermal contrast is maximum. In order to obtain the contrast image C_{img}, contrast computations are done on the whole image (all the rows). All the images of the time zone k_{max} where thermal contrast is maximum are selected and Eq. (6.47) is applied to each pixel $T(m, i, j)$ of the mth image (converted into temperature) of time zone k_{max} (i = row, j = column in the image):
For $\forall j, j = 0, 1, (\text{Maxcol-1})$, for $\forall i, i = 0, 1, \ldots, (\text{Maxrow})$:

$$C_{img}(k_{max}, i, j) = \frac{1}{N_{k_{max}}} \sum_{m=0}^{N_{k_{max}} - 1} \frac{T(m, i, j) - T0(i, j)}{\frac{1}{s} \sum_{l=0}^{l=s-1} [r\_far\, T(m, l) - r\_far\, T0(l)]} \qquad (6.50)$$

where $N_{k_{max}}$ = number of images in time zone k_{max}.

For example, for the case of Fig. 6.9, we give in Fig. 6.11 (see colour section) three images: (a) raw image (one taken among images of zone k_{max}), (b) the (unsmoothed) contrast image C_{img} obtained using Eq. (6.50) and (c) the (smoothed) contrast image C_{img}, i.e. image (b) smoothed using the sliding Gaussian of Sect. 3.2.2. We notice the smoothed (maximum) contrast image C_{img} is well corrected (uniform background, good defect visibility) and ready for defect shape extraction as we will see in the next section. This better appearance of contrast image was foreseeable since this image is obtained after enhancement procedures which reduce radiometric distortions (through temperature computations) and eliminate parasitic effects of thermal reflection and emissivity.

Conventional enhancement procedures used for visible images are less effective than the procedure presented here. On the other hand, these procedures are faster since they are generally local in nature and operate on a single image basis. For instance, Lee (1980) proposes processing pixels in a small window (of 3×3, 5×5 or 7×7 pixels). The centre window pixel p is replaced by p':

$$p' = m(1 - k_1) + k_1 p$$

where k_1 is a gain factor and m is the average window value. If $k_1 = 1$, no effect is obtained, if $k_1 = 0$ the pixel is replaced by the average of the window (this smooths the image), if $(0 < k_1 < 1)$ the effect corresponds to a low-pass filtering and with $(k_1 > 1)$ contrast will be enhanced by increasing the effect of high frequency content of the image (high-pass filtering). The following references describe other enhancement algorithms: Gregory (1982); Hughett (1987); O'Gorman (1988); Beghdadi and Le Negrate (1989); Frock (1989); Leszczynski and Shalev (1989); Pennington and Moorhead (1990).

Finally, notice that obviously the technique presented in this section will be more accurate if q (q is the image acquisition rate; i.e. number of thermal images recorded per second) is high.

6.3.4 Defect Shape Extraction

Gradient Computation

We recall that in Sect. 4.4.4, we presented an extraction algorithm which permits us to determine defect shape; this algorithm was based on an adjustable threshold gradually lowered around the hottest pixel of the defect (defect centre). This method was practical and fast, but was only an approximation. As we saw in Sect. 6.2, the exact defect shape can be obtained from the apparent shape which corresponds to the steeper temperature gradient computed on the sample surface over the detected defect. In this section, we will discuss gradient computation and defect shape extraction. The gradient will be computed from the smoothed maximum contrast image C_{img} (as in Fig. 6.11c, d).

It is necessary to differentiate the maximum contrast image (Eq. (6.50)), once smoothed (Sect. 3.2.2), in order to obtain the gradient image. One of the simplest differentiation methods consists of computing the Roberts gradient, which is defined with respect to the cross difference of Fig. 6.12a and which is defined as follows (Gonzalez and Wintz 1987, p 177):

$$|G| = G[f(i, j)] = \{[f(i, j) - f(i+1, j+1)]^2 + [f(i+1, j) - f(i, j+1)]^2\}^{1/2} \quad (6.51)$$

In this formula f is the image on which the gradient $|g|$ is computed and i, j are image indexes. As we will see below, this first order approximation produces noisy gradient images from which defect extraction is more delicate. Consequently, it is better to rely on an higher degree of approximation.

In App. D, a higher order gradient approximation is derived. Using this method, the gradient image $|G|$ is computed by means of the following formula applied in a

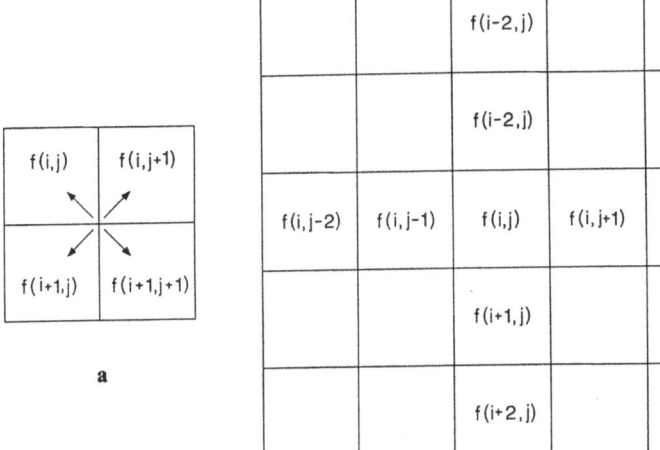

Fig. 6.12. Pixel relations for **a** Roberts gradient; **b** gradient computed using Eq. (6.52).

cross fashion on both image f rows and columns as shown in Fig. 6.12b (Eq. (D.4)):

For $\forall i, i = 2, 3, \ldots, \text{Maxcol-3}$

and for $\forall j, j = 2, 3, \ldots, \text{Maxrow-3}$, compute

$$|G| = \left\{ \begin{array}{c} [f(i, j-2) - 8f(i, j-1) + 8f(i, j+1) - f(i, j+2)]^2 \\ + [f(i-2, j) - 8f(i-1, j) + 8f(i+1, j) - f(i+2, j)]^2 \end{array} \right\}^{1/2} \qquad (6.52)$$

and the edge pixels are obtained through duplication. This method is more accurate than the Roberts gradient (Eq. (6.51)) without requiring much more time for computation (see below). Sensitivity to residual noise present in the smoothed contrast image is one of the main disadvantages of this method.

If a smoother gradient image is wanted, another method can be used which is based on a second order fit over the few pixels taken in a cross fashion around the pixel of interest at location indexes (i, j) in image f, Fig. 6.12b (see App. C for the fitting computation). For each direction (row i, column j), the second order fit function can be expressed by

$$f_r(i) = a_r + b_r i + c_r i^2$$
$$f_c(j) = a_c + b_c j + c_c j^2$$

where a_r, b_r, c_r and a_c, b_c, c_c are the parameters found by the fitting procedure (along the row and column respectively). Note that to avoid numerical errors in the fitting procedure for pixels of high index, it is necessary to replace the exact index location by $-3, -2, -1, 0, 1, 2, 3$, where 0 represents pixel (i, j). The first derivative is then given by

$$f'_r(i) = b_r + 2c_r i \sim b_r$$
$$f'_c(j) = b_c + 2c_c j \sim b_c$$

and then $|G|$ is given by

$$|G(i, j)| = [f'_r(i)^2 + f'_c(j)^2]^{1/2} \qquad (6.53)$$

Table 6.3.

Method to obtain gradient image $|G|$	Relative computation time
Eq. (6.51), Roberts gradient	1
Eq. (6.52)	1.5
Eq. (6.53), 3×3 cross	8
Eq. (6.53), 5×5 cross	11
Eq. (6.53), 7×7 cross	12
Eq. (6.53), 11×11 cross	16

Figure 6.13 (see colour section) presents typical results for the gradient computation. As expected, the Roberts gradient (Eq. (6.51)) exhibits a strong noise level. Higher degree approximation (Eq. (6.52)) shows improvement in this respect (labelled "ROBERTS_ 2nd" in Fig. 6.13). Better results are obtained if the method of Eq. (6.53) is applied (labelled "FIT_2" in Fig. 6.13), particularly if the fitting procedure spans more pixels, such as when using an 11×11 kernel.

Concerning processing time, however, the method of Eq. (6.53) is costly (Table 6.3).

Defect Shape Extraction

From the gradient image $|G|$, the next step is to extract the defect shape which corresponds, to a given factor (Sect. 6.2.2), to the locus of maximum derivative around defect centre located in i_d, j_d (centre of the crater in Fig. 6.13). This operation can be performed using a simple technique. From the defect centre, a line segment D of orientation θ is rotated $0 < \theta < 360°$ and the values (posx, posy) corresponding to maximum gradient along D are found (Fig. 6.14):

$$\text{pos}y = j_d - (\text{pos}x - i_d)\tan\theta \tag{6.54}$$

For a given θ, posx is incremented by 1 at each iteration (posx $= i_d, i_d + 1, i_d + 2, \dots$) until the maximum $|G|$ (posx, posy) is found through a search done within all the values obtained along D. The algorithm presented in Sect. 4.4.4 can be easily adapted

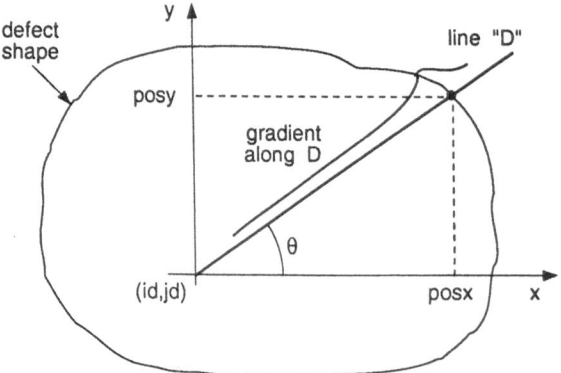

Fig. 6.14. Method used for contour extraction of the defect (see text).

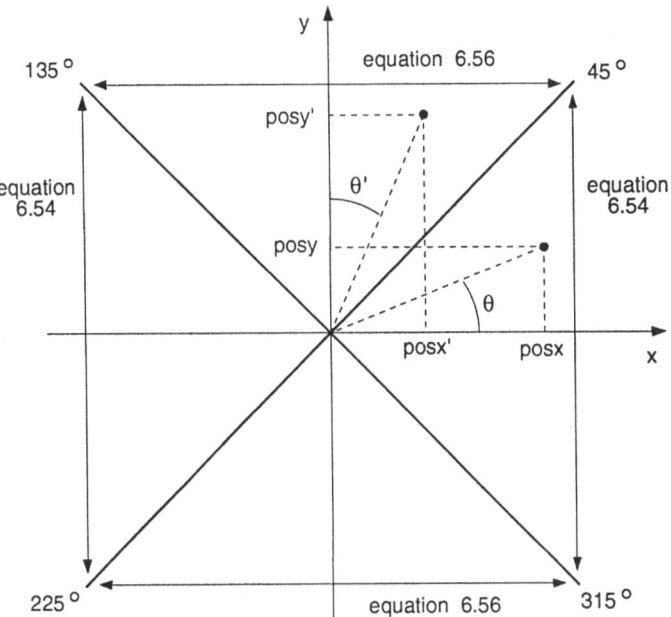

Fig. 6.15. Illustration of Eqs (6.54) and (6.56).

to perform this maximum search successfully (remember that since the differentiation process increases noise content, gradient images are more delicate to work with; a simple comparison between Figs. 6.11 and 6.13 shows this clearly). We proceed in this fashion by rotating the line around the defect centre. At the end of this process, we obtain the list of coordinates for vertexes of defect shape:

$$\text{posx }[i] \text{ and posy }[j] \tag{6.55}$$

where $i = 0, 1, 2, P-1$

$$P = \text{number of angular position around } (i_d, j_d)$$

To avoid problems related to infinite values of $\tan 90°$ in Eq. (6.54), we use Eq. (6.54) for $0° < \theta < 45°$, $135° < \theta < 225°$, $315° < \theta < 360°$. For orientations $45° < \theta < 135°$ and $225° < \theta < 315°$, we work with θ shifted $90°$ with respect to θ (Fig. 6.15):

$$\text{posx}' = i_d - (j_d - \text{posy}) \tan \theta' \tag{6.56}$$

The number of angular steps P is specified by an initial maximum gradient search at $\theta_1 = 0°$, $\theta_2 = 90°$, $\theta_3 = 180°$, $\theta_4 = 270°$ in gradient image $|G|$ in order to find the number of columns (or rows) available for the increment of posx (or posy') so that no point lies on the same column or row as its neighbour (Fig. 6.16):

$$G_{\text{max},\theta=0°} \Leftrightarrow (i_d, j_{\text{max}\_\theta=0°})$$

$$P_0 = (j_{\text{max}\_\theta=0°} - j_d)$$

and P_{90}, P_{180}, P_{270} are computed in the same way. The maximum value is next selected along rows and columns: $P_1 = \max(P_0, P_{90}, P_{180}, P_{270})$ and

$$P = 4P_1 \tag{6.57}$$

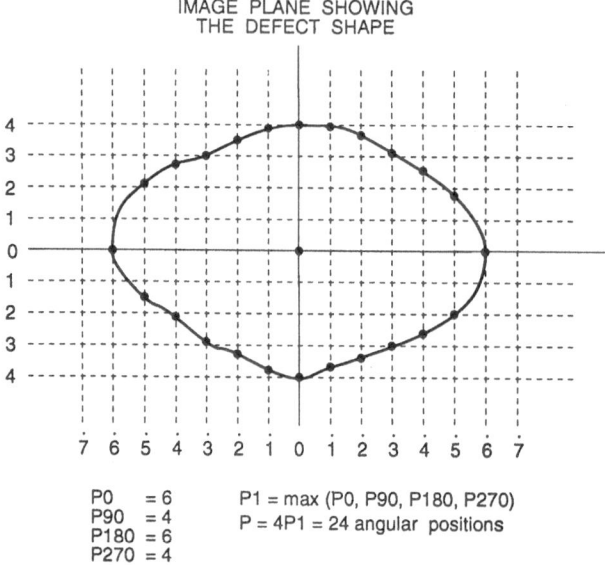

IMAGE PLANE SHOWING
THE DEFECT SHAPE

PO   = 6   P1 = max (P0, P90, P180, P270)
P90   = 4   P = 4P1 = 24 angular positions
P180 = 6
P270 = 4

Fig. 6.16. Illustration of the computations needed to establish the number of steps which define the defect contour.

This method of extracting the apparent defect contour by finding the list of vertexes (Eq. (6.55)) is advantageous since it allows us to correct easily the distortion effects of image fields not having the same number of rows and columns, as we will see now. A practical "trick" to measure the field of view is to place a warm bright metallic ruler with black painted lettering at the focus of the infrared camera's field of view. Due to difference of emissivities, the lettering is seen with a good contrast with respect to the bright metal.

Without considering the LSF or SRF (Sect. 3.1.1), it is possible to compute a geometric factor which relates pixels with apparent size in the field of view and also compensate for the unequal number of rows and columns.

For instance, suppose

$$105 \text{ columns} = \text{Maxcol, horizontal field size} = 32.9 \text{ mm}$$

$$68 \text{ rows} = \text{Maxrow, vertical field size} = 35.9 \text{ mm}$$

Thus

$$\frac{32.9 \text{ mm}}{105 \text{ pixels}} = 0.313 \text{ mm/pixel along columns } (x)$$

$$\frac{35.9 \text{ mm}}{68 \text{ pixels}} = 0.528 \text{ mm/pixel along rows } (y)$$

In this example, there is a distortion of the field of view since the *apparent* size of a pixel is different along rows and columns. For instance, if a square object placed in the field of view fills a 20×12 pixel surface on the infrared monitor screen:

$$\text{Along } x \rightarrow 20 \text{ pixels represents } 6.26 \text{ mm}$$

$$\text{Along } y \rightarrow 12 \text{ pixels represents } 6.34 \text{ mm}$$

If we multiply dimensions along y by factor $f_y = 0.528/0.313 = 1.687$ to display the

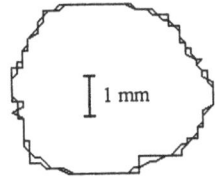

Fig. 6.17. Example of Fig. 6.9, extraction of apparent defect contour d_{app}. The shape is shown raw and smoothed. In this case Eq. (6.52) was employed to compute the gradient, thus explaining the noisy shape obtained prior to smoothing. Graphite epoxy anisotropy is visible.

object without apparent spatial distortion then

$$\text{Along } x \rightarrow 20 \text{ pixels represents } 6.26 \text{ mm}$$
$$\Rightarrow 6.26/20 = 0.313 \text{ mm/pixel}$$
$$\text{Along } y \rightarrow 12 \times 1.687 = 20.24 \text{ pixels represents } 6.34 \text{ mm}$$
$$\Rightarrow 6.34/20.24 = 0.313 \text{ mm/pixel}$$

and the object is displayed correctly as a square on the screen (the image displayed is of equal size in x and y); f_y can thus be expressed by

$$f_y = \frac{\text{Maxcol} \times \text{vertical field}}{\text{Maxrow} \times \text{horizontal field}} \tag{6.58}$$

The method based on a rotating line segment to extract defect shape allows easy correction of the unequal vertical and horizontal image fields since it is only necessary to multiply all values of posy $[i]$ of the vertex list (Eq. (6.55)) by f_y (Eq. (6.58)).

For instance, Fig. 6.17 shows the results of defect shape extraction performed with methods reviewed in this section (this is the example of Fig. 6.10). Superimposed on the defect contour is a smoothed plot obtained using closed spline functions (complete derivation of this smoothing algorithm can be found in Laurendeau (1982, p 62ff)). Median filtering or sliding Gaussian noise smoothing techniques discussed in Sect. 3.2.2 can also be applied on computed radii between the point of interest (posx $[i]$, posy $[i]$) and defect centre (i_d, j_d) to correct for uneven shapes. The difficulty is to be certain to obtain a closed shape with no steep transition at the extremity (two smoothing passes with different starting points around the defect centre can solve the difficulty).

In this section, we have studied the different operations required to obtain information needed to characterize quantitatively detected subsurface defects (the detection step proceeds as explained in Chap. 4). These operations concern the preparation of the acquisition of a thermogram sequence (particularly the time scale) and also the subsequent computation. Once this experimental information about the inspected sample is obtained, the inverse problem can be tackled, as explained in Sect. 6.2. In the next section, we will briefly discuss the expected accuracy of procedures reviewed in Sects 6.2 and 6.3.

6.4 Discussion on the Quantitative Characterization Procedure

In this section, we will introduce some elements of discussion which are of interest when a quantitative inspection analysis is performed. One potential difficulty is to select a sound area within an image (vector r_far, Eq. (6.47)). This has implications, due to the three-dimensional spreading of the heat front. In particular, contrast

Table 6.4.

Evaluated parameter	$\%^a$	$\%^b$
Depth	8.1	13.8
Thermal resistance	24	126
Size	4.6	NA

[a] Krapez et al. (1991).
[b] Balageas et al. (1987c).
NA: Not available since the analysis is one-dimensional.

computation can be affected, corrupting the quantitative evaluation. In fact, since contrast computation is performed relatively between the zone of interest (above detected defects) and a sound area so that a unit contrast is obtained in the absence of defect (Eq. (6.28)), thermal contrast values will be corrupted if the "sound area" is selected too close to the defect and is thus submitted to its influence (computed thermal contrast will be of smaller amplitude). This effect will be more evident if thermal images are of small size.

In order to give an idea of the expected accuracy of quantitative inversion procedures discussed in this chapter (or of similar TNDE inversion procedures found in the literature), we present now some typical results. In the case of graphite epoxy samples with inclusion-type defects made by inserting Teflon™ film (100 µm thick) between two plies of prepreg), Balageas et al. (1987c) and Krapez et al. (1991) obtained the figures shown in Table 6.4, using the thermal stimulation approach in reflection. The numbers indicated correspond to the percentage difference between defect nominal values and defect measured values using the inverse method. One of the major causes of discrepancy in these values lies in accounting for thermal losses which take place during the experiments between the sample and the surroundings. These losses cannot be ignored if the experiment lasts longer than a few seconds. They are, however, difficult to handle in the inverse problem, especially if the erratic behaviour of air convection phenomena which take place during thermal inspection experiments has to be taken into account.

As seen from the numbers in Table 6.4, the expected accuracy of inverse analysis performed with experimental data is a few percent for depth and size evaluation, but worst for thermal resistance (as expected from Fig. 6.5). One reason for this is that during the preparation of samples, air can be included with the artificial defect inclusion. This can be the case if Teflon™ implants are inserted between plies of graphite epoxy prepreg in order to simulate an inclusion. The presence of air layers not accounted for perturbs the inversion procedure. Many authors have noticed the presence of air layers in samples after destructive testing of such CFRP specimens.

As a last consideration, it is worthwhile to mention the necessity to perform digitization of thermal images using a sufficient number of bits. In the past, systems were offered using 6 or 7 bits. Currently, 8 bits are used while 12 bit digitization is emerging seriously. Such a high level of digitization allows us to reduce noise level at the acquisition stage thus permitting capture of small thermal events and better agreement of the inversion procedure with real data.

Conclusion. In this chapter, we have studied methods which allow us to extract quantitative information from TNDE data. Both the inverse problem and experimental image processing techniques were reviewed. The expected accuracy of quantitative analysis is around a few percent.

Chapter 7

Inspection of Materials with Low Emissivity by Thermal Transfer Imaging

7.1 General Considerations

Application of TNDE is difficult, if not impossible, in the case of materials having a low emissivity. It can be shown that radiation measurement cannot be successfully performed if material emissivity is smaller than 20% of black-body emissivity (Gaussorgues 1984b, p 421). This is, for instance, the case for materials having highly reflective surfaces, such as bare metal sheets. The bonded aluminium laminate samples studied in Chaps 1 and 4 and which are used in aeronautic and transport industries are good examples of such materials.

Metal surfaces have a small absorptivity and thus a weak emissivity in the infrared, about 5% of the black-body emissivity (Sect. 2.1). Consequently, radiance emitted by such metallic surfaces is weak and produces only faint infrared images. Worse, presence of grease patches or slightly oxidized zones can change surface emissivity by factors of 5%, 10% or 20% of the black-body emissivity. This yields *apparent hot spots* in the infrared images, hot spots which can be falsely interpreted as damaged zones following active thermography principles discussed previously. Moreover, high reflectivity of metal surfaces introduces a problem of parasitic reflection, where emitted radiation reflected by warm surrounding bodies is reflected by the metal surface, which acts as a mirror in the infrared spectrum. This reflected image superimposes with the image emitted by the metal surface under inspection, creating parasitic hot spots which further complicate thermogram interpretation. Equation (2.6) puts this question in evidence. An illustration of the problem is shown in Fig. 7.1 (see colour section). An infrared camera is pointed directly at a shiny aluminium sheet and instead of revealing the uniform temperature distribution over the sheet, the image reveals the isotherm distribution of the infrared camera operator reflected by the sheet. Consequently, extreme care is required when TNDE is to be employed for materials having reflective surfaces. Note that in the case of natural outdoor scenes, emissivity is generally high, with the exception of water or ice surfaces observed at low incidence (< 40°) viewing angles (Stillwell 1981).

Black-Painting

A possible approach to solving the emissivity problem consists of covering metallic surfaces before the inspection with a black paint. Such paints have uniform and high emissivity ($\varepsilon \sim 0.9$) in the infrared bands, and low reflectivity (Eq. (2.4)). Consequently, if they are applied to low emissivity surfaces, interference caused by neighbouring warm bodies is reduced and thermographic inspection becomes possible. The necessity to "black-paint" prior to inspection and to remove the paint afterwards is, however, not attractive for an NDE technique which claims a fast inspection rate as one of its strongest advantages with respect to other inspection techniques (Sect. 1.5). Moreover, black-painting components can also modify radiometric readings. For instance, if paint is applied to PCBs, temperature distribution over the inspected components can be altered due to the enhanced heat dissipation (Dumpert 1992). Emissivity depends on the paint coating thickness; tests have shown that if three or more layers of paint are applied, the emissivity becomes constant (McLaughlin 1988; Balageas et al. 1988).

Techniques to Solve the Emissivity Problem

Considering the emissivity problem, Vleck (cited by Öhman 1981) proposed an original technique well suited to the case of small specimens whose emissivity is unknown or uneven over the object surface. His technique consists of keeping the specimen between clasped hands long enough to ensure the specimen reaches skin temperature (this can be checked with a small temperature probe inserted between the sample and one of the palms). Once the specimen temperature has stabilized, the hands are opened and an image of both palms and specimen recorded immediately with the infrared equipment. Since skin emissivity is high ($\varepsilon \sim 0.97$), the emissivity distribution over the specimen surface can be determined by comparing temperature distributions recorded over both the palms and the specimen which, due to this primitive thermal transfer imaging technique, are the same (if the exposure proceeds quickly enough). Of course the object emissivity must be high enough to avoid the problem of thermal reflections.

Another possible approach to solving the emissivity problem is to use a reflecting cavity (Cielo et al. 1988a; Chen et al. 1990; Krapez et al. 1990; Cielo 1992). This technique is appropriate when it is necessary to obtain an average temperature reading above the surface; it is, however, not suitable for thermal imagery since multiple reflections tend to average the radiation emitted by the surface covered by the cavity. Such an approach eliminates the spatial resolution needed for localization and characterization of damaged zones.

Other techniques are possible, but they require more complex experimental apparatus. For instance, it is possible to measure directly the surface reflectivity using a laser reflectometer prior to taking temperature readings (Green 1968; Peacock 1988). Other techniques use two temperature sensors. For instance, in the steel industry, to determine oven temperature, one sensor is pointed at the steel surface while the other is pointed at the oven wall in order to take into account spurious thermal reflections (Johnson et al. 1988; Ramelot et al. 1988). Note that these methods are not useful if contributions from spurious reflections are much more important than those of interest (Roney 1982). Another drawback is that these techniques have generally a limited spatial resolution (point mesurement only).

Multiwavelength pyrometry can also be used. However, this technique is generally restricted to high temperature (Reynolds 1964; Coates 1981; Hunter et al. 1985, 1986; DeWitt, 1986; Braim and Cuthbertson 1986; Auric et al. 1987; Brownson et al. 1987; Johnson et al. 1988; Maldague and Dufour 1989; Mansoor et al. 1991 a,b). Multiwavelength pyrometry is based on the assumption that a simple relationship between emissivity and wavelength exists, so that it becomes possible to calculate the temperature from measurements taken at different wavelengths (it is like solving a set of N equations having N unknowns, with temperature and emissivity being unknowns). For two-colour pyrometers temperature is computed from the ratio R of the two signals recorded at both wavelengths since, for a given set of wavelengths λ_1, λ_2 and temperature T, the spectral radiance ratio is unique (Fig. 2.2). In the case of two-colour pyrometry the hypothesis is that the emissivity ratio is constant over the wavelength span of interest. Experimental measurement of the signal ratio R corresponds to a temperature value $T = f(R)$ after proper calibration where $f(\cdots)$ is the calibration polynomial. The assumption of constant emissivity ratio holds for *grey bodies* (for which emissivity is constant with wavelength) and to some extent to coloured bodies for which emissivity varies with wavelength), consequently the calibration function $f(\cdots)$ of the pyrometer is specific for a given material.

In the next section, we review some techniques which solve this emissivity problem by using thermal transfer imaging. The principle of thermal transfer imaging is based on a high emissivity material continuously brought into contact with the structure under inspection. Since the thermal observation is performed on the high emissivity material which picks up the temperature distribution of the low emissivity material it is in contact with, TNDE inspection becomes possible. Such a technique is suitable for low temperature inspection.

7.2 Thermal Transfer Imaging

In the previous chapters, we showed how infrared thermography allows detection, localization and characterization of subsurface flaws in materials. In order to concentrate on fundamental concepts, the question of emissivity was not really reviewed. All the inspected samples whose inspection results were presented in these chapters were systematically black-painted (with the same paint as the one used during the calibration process). The thermal transfer imaging technique we now discuss is restricted to the inspection of flat reflective surface components such as aluminium sandwich panels (cf. Sect. 4.3).

Using this transfer technique, the emissivity problem is solved by placing, along the inspection line, a high emissivity freely rotating flexible roller (Cielo et al. 1991; Maldague et al. 1991b). The roller is positioned in such a manner that it is continuously in good thermal contact with the (low emissivity) surface under inspection. The thermal front propagates from the surface to the roller so that, after contact, the temperature distribution of the roller surface reproduces the temperature distribution of the surface. This transferred temperature distribution can be observed by pointing the infrared camera directly on the flexible roller surface as shown in Fig. 7.2. Since the roller surface is of high emissivity and thus of low reflectivity (Eq. (2.4)), the thermal image picked up on the roller has a relatively good thermal contrast and is not affected by parasitic reflections.

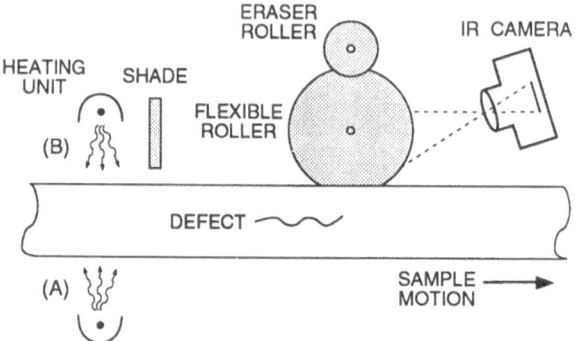

Fig. 7.2. Thermal transfer imaging system using a compliant roller and shown in **A** a transmissive and **B** a reflective configuration for the detection of defects of known depth in a laminar specimen.

The flexible roller can be made of polymer foam coated with a high emissivity paint; the large number of gas bubbles ensure good contact and low thermal conductivity. In the region where the roller is in contact with the inspected surface, the gas bubbles collapse due to the pressure; this ensures good transfer of the thermal front. After contact, the bubbles expand back to their initial volume. Since foam thermal conductivity is weak, the transferred image, observed with the infrared camera, offers good thermal contrast.

An auxiliary roller, preferably made of a high thermal conductivity metal and cooled using internal water circulation or Peltier elements, is also brought into contact with the flexible roller. It acts as an "eraser" for the transferred thermal image.

The thermal perturbation can be deployed either in reflection (inspection of thick components) or in transmission (inspection of thin components). Since the emissivity of the surface under inspection is small, instead of using a radiative type of thermal perturbation source (such as line heating), a contact thermal stimulation source such as water or air jets is preferred. Note, however, that if line heating is deployed in conjunction with a back reflector, the effective emissivity is relatively high, due to multiple thermal reflections between the parabolic reflector and the surface, thus making the energy deposition possible (in some instances).

In the case of shallow defects, the thermal perturbation unit can be suppressed by maintaining the auxiliary eraser roller at a sufficiently different temperature with respect to the surface under inspection. In this case, the thermal perturbation is applied to the surface during contact with the flexible roller and the response to this perturbation is immediately collected by the roller. Note also that the infrared camera can be replaced by an infrared line scanner.

Another configuration based on the principle of the set-up of Fig. 7.2 is shown in Fig. 7.3. In this case a closed-loop membrane connects both rollers, allowing the circumference of the transfer head to be reduced. This allows the collection of short thermal contrasts which would be lost if thermal transfer occurred later (case of shallow subsurface flaws or inspected material with high thermal conductivity). An additional benefit of this configuration is the ease with which the infrared image can be observed and subsequently erased, due to the membrane length.

In the case of the configurations shown in Figs 7.2 and 7.3, subsurface defects must have a depth of the order of the thermal propagation length corresponding to the period of time between injection of the perturbation and transfer of the temperature

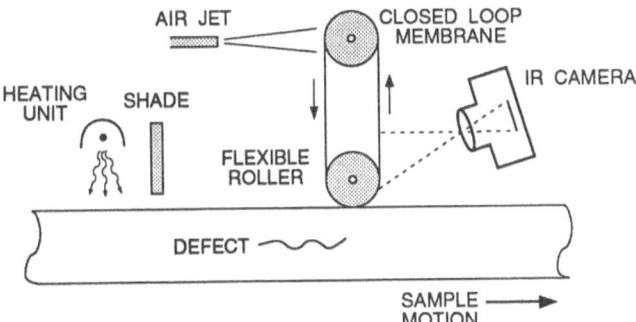

Fig. 7.3. Thermal transfer imaging using the same basic principles as the configuration of Fig. 7.2 and shown for the reflection approach, with a closed-loop membrane for reduced head circumference and more convenient erasure of the transferred images.

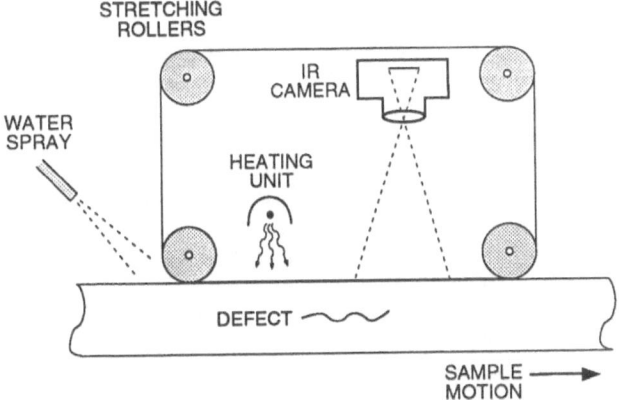

Fig. 7.4. Thermal transfer imaging system using a closed-loop membrane applied on the workpiece surface for the detection of defects of unknown depth.

distribution by the roller. Consequently, the distance between the thermal source and the roller must be selected according to the depths of the possible defects (for instance the thickness of the bonded sheet in sandwich assemblies; see Sect. 1.2.2).

For specimens having a wide range of possible defect depths, the modified configuration of Fig. 7.4 can be employed. In this case, a high emissivity membrane made of a thin film is brought into contact with the panel surface using stretching rollers. Tiny water sprays are used to produce, by capillary action, a thin film of liquid between the membrane and the surface under inspection in order to obtain good thermal contact and also to hold the membrane in place. Active thermography is employed following methods discussed in Chap. 4. The thermal front is eventually affected by the presence of subsurface flaws, thus causing hot spots or cold spots (depending on the nature of the thermal perturbation) to appear on the membrane surface observed by the infrared camera.

Modifications can be introduced to the configuration of Fig. 7.4. For instance, the thermal perturbation can be applied by water sprays or by the first stretching roller if a sufficiently large temperature differential is maintained between this roller and the surface under inspection.

In the case of nonplanar surfaces, it is possible to modify the configuration of Fig. 7.4. A high emissivity flexible membrane can be made to follow the specimen contour either by pressing it on the surface (a thin film of water helps to hold the membrane in place) or by means of a vacuum sealing machine. Following this operation, the TNDE inspection can proceed as if the surface has been black-painted using the techniques described in preceding chapters.

Finally, note that thermal contrasts observed by these transfer methods are smaller with respect to contrasts obtained by direct observation (with black-painting, see next section for more details).

7.3 Physical Behaviour of Thermal Transfer Imaging Technique

It is possible to simulate numerically the behaviour of the thermal transfer process in order to understand the underlying physical principles better. An adapted version of the model presented in Sect. 2.2.2 can be used for this purpose. In this section, we will restrict this discussion to some meaningful results.

The thermal process can be divided into three phases (Fig. 7.5). In the first phase the surface of the sample is heated and cools down. In the second phase, from time t_1 to t_2, the flexible roller is compressed and is in contact with the surface under inspection. In the third phase, at time t_2, both materials are separated and the flexible roller re-establishes itself. Note that instead of using a hot thermal perturbation source, cold thermal perturbation can be used. In this case the sample surface starts to warm just after the cold thermal perturbation is over.

Figure 7.6 presents some interesting results. Note that on the abscissa the normalized time is expressed in Fourier numbers (Eq. (6.11)); this permits us to derive more general results. In the figure, the second phase starts at Fourier number $F_{phase2} = 1$; this time corresponds to a thermal propagation time of the thermal front to the depth z_{def} of the defect of the order of $t = z_{def}^2/\alpha$. The various results considered in Fig. 7.6 are for different durations of the second phase (period of contact between foam and specimen), $F_{phase2} = 1, 3.2, 10, 32$ with $F_{phase3} = 10$. Two situations are investigated: (a) for materials of high thermal conductivity such as aluminium (case

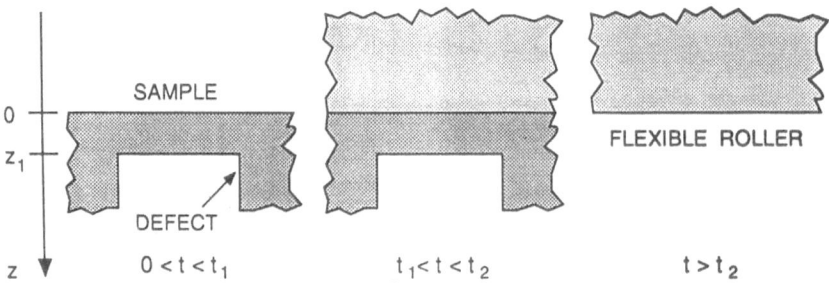

Fig. 7.5. The three phases of the thermal transfer process.

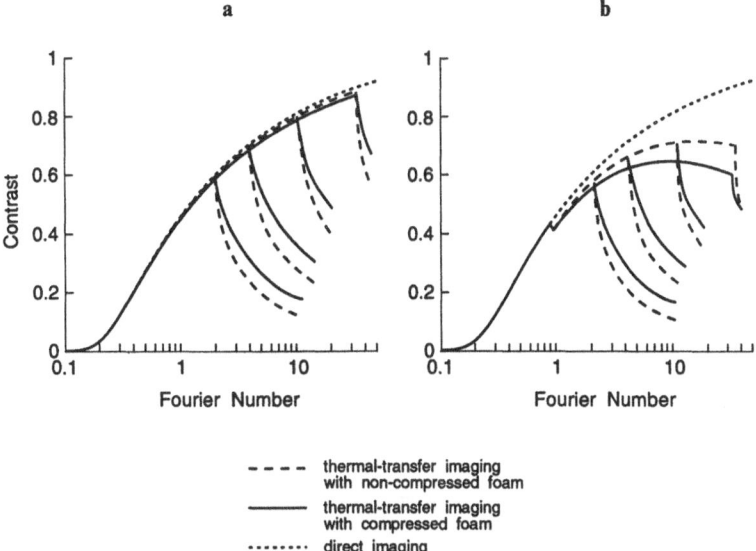

Fig. 7.6. Contrasts computed with different imaging methods for materials of **a** high and **b** low thermal conductivity.

illustrated) and (b) for materials of low thermal conductivity such as graphite epoxy (case illustrated). The contrast obtained by observing the surface under inspection directly is shown with a dotted line (in this case, the computer model assumes the surface is covered with a high emissivity paint prior to inspection). Other cases are shown corresponding to thermal transfer imaging observation in the case of both uncompressed and compressed foam. Notice that the results for this model are less valid for large time values (Fourier number > 10) since the model used to draw Fig. 7.6 is one-dimensional and does not consider thermal losses. This explains why, in this figure, the contrast increases steadily. In a real situation, the contrast in direct observation would reach its maximum around Fo ~ 1 to 10 before decreasing because of thermal losses (e.g. Fig. 6.10).

Nevertheless, study of Fig. 7.6 reveals interesting aspects: thermal transfer imaging is less efficient than direct observation in all but in the case of low emissivity surfaces. In this latter case, images must be transferred as soon as possible after the separation since the temperature drops abruptly thereafter. Consequently, the configuration of Fig. 7.3 is more useful than that of Fig. 7.2 because of its smaller head circumference. Uncompressed foam seems also to give higher contrasts, but this advantage disappears rapidly after separation. At longer times, compressed foam is more suitable for acquisition of image sequences (cf. Sect. 6.3.2), since it offers higher contrast.

7.4 Experimental Results

In this section we will illustrate the principles of operation of the thermal transfer imaging technique with a few experimental examples. Two kinds of materials of high and low thermal conductivity were tested: first, an aluminium–foam bonded panel,

Fig. 7.7. Experimental apparatus using the configuration of Fig. 7.2: the infrared camera, flexible black-painted roller and lamp heater, as well as the aluminium laminate mounted on a slide (extreme right) are visible.

similar to those specimens studied in Chap. 4. An artificial bonding defect was created: a lack of adhesive over a 4 cm diameter circular area, 1 mm below the surface (aluminium sheet thickness = 1 mm). A Plexiglass™ plate, in which a 1 cm diameter hole was drilled 1.6 mm deep, was also studied. Samples were inspected in reflection using a 1600 W line heater.

Figure 7.7 shows the experimental apparatus with the flexible foam roller. Figure 7.8 (see colour section) presents a few results obtained for both the aluminium panel and the Plexiglass™ plate. Three images are shown: on the left the black-painted aluminium panel was tested in reflection with the camera directly pointed at the painted surface, while the panel mounted on a slide moved at about $10 \, \text{cm} \, \text{s}^{-1}$ in the field of view. Thanks to the high emissivity of the paint ($\varepsilon \sim 0.9$), the debonding defect is clearly visible in the middle of the thermogram. A test performed under the same conditions on the same panel, unpainted, did not reveal the defect because of the low emissivity of the surface.

The image on the right of Fig. 7.8 shows the same panel, but observed using the thermal transfer imaging set-up of Fig 7.7, where the camera was pointed at the surface of the flexible foam roller. Although the thermal contrast is smaller than in the case of direct observation, the presence of the defect is still visible. Note the aluminium surface was black-painted in order to be able to stimulate sufficiently the specimen with the radiative heat source. The hotter band at the bottom of the thermogram corresponds to the observation of the black-painted panel surface, hotter than the roller and observed by the camera.

The image at the bottom of Fig. 7.8 shows a thermogram obtained for the Plexiglass™ plate. The transferred image reveals clearly the presence of the subsurface hole (infrared camera pointed on roller surface). The plate surface was black-painted in order to absorb enough energy from the radiative stimulation device. For these low emissivity surfaces, radiative heating is not very efficient; contact thermal stimulation devices using either water or air jets are more appropriate, as discussed in Chap. 4.

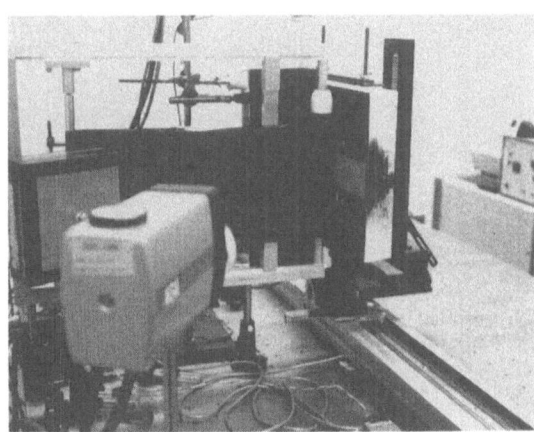

Fig. 7.9. Experimental apparatus using the configuration of Fig. 7.3: the infrared camera, flexible black-painted roller, closed-loop membrane and lamp heater are visible as well as an aluminium laminate with foam core mounted on a slide. Notice the unpainted aluminium surface being inspected.

Figure 7.9 shows the experimental apparatus for configuration of Fig. 7.3 in which a rubber membrane 1.5 mm thick is used to catch the thermal images. Figure 7.10 (see colour section) shows a reconstructed image (using the techniques of Chap. 4). The total circumference of the inspection head is of the order of 10 cm, much less than the 30 cm or so of the set-up of Fig. 7.7. This allows thermal contrasts to be captured earlier than with the larger roller (cf. Fig. 4.9b), thus enabling observation of the subsurface bonding defect in the aluminium panel. A pressurized air curtain is used to erase thermal imprints on the membrane (box visible on the left of Fig. 7.9).

Figure 7.11 (see colour section) shows results obtained for the configuration of Fig. 7.4. A thin high emissivity membrane is brought into contact with the high reflectivity aluminium surface in order to increase its emissivity. For the test of Fig. 7.11, a 50 μm film of black polyurethane is brought into contact through capillary action with the metal surface, as explained previously. A segmented image (Sect. 4.4) is shown in Fig. 7.11a. Fig. 7.11b presents a segmented image of the same bonding defect but in direct observation (with the aluminium surface black-painted and without using the polyurethane film). A comparison between the two images in (Fig. 7.11) reveals a loss of contrast (Fig. 7.11a); the defect is, however, visible in both cases. For this configuration (Figs 7.4 and 7.11a), since the complete temporal evolution of temperature recorded above the defect is available, characterization techniques of Chap. 6 can be employed (with adjustments to compensate for the loss of contrast); this is not the case of the other two configurations where observation is performed instantaneously.

Conclusion. In this chapter, we have presented an approach which allows the use of TNDE in the case of high reflectivity (flat) components. This approach enables us to solve, in certain circumstances, the important emissivity problem of TNDE (and its correlated problem of parasitic reflections).

Chapter 8

Thermal Diffusivity Measurements of Materials

8.1 General Considerations

Evaluation of material properties is a field of great interest for TNDE (Imhof et al. 1991). This is an important subject for at least two reasons. First of all, in order to predict the behaviour of a given component or solve the "inverse problem", it is necessary to perform modelling of the actual experiments. Such modelling requires knowledge of thermophysical properties of the inspected materials.

Measurement of thermophysical material properties, on-site, is thus advantageous since it allows better simulation of thermal behaviour of the components under inspection. Values found in tables and handbooks such as the remarkable *Thermophysical Properties of Matter* (Touloukian and DeWitt 1970) give only an estimation of such values since material properties may change due to the variability of the fabrication processes. A more accurate figure is obtained by direct evaluation.

This brings us to another motivation for determination of material properties. In fact, it can be shown that in some instances, measurement of one thermophysical property is closely related to the value of another material property. For example, Peralta et al. (1991) indicated there are consistent differences in thermal diffusivity of high purity aluminium specimens depending on the degree of specimen recrystallization.

In the same way, Heath and Winfree (1989) showed that in the through-ply direction, a roughly linear functional dependence exists between porosity and thermal diffusivity in carbon–carbon composites. This is an interesting result, since porosity of such components is reduced through repetitive processing cycles (a processing cycle consists of immersion in phenolic resin followed by pyrolysation: the resin penetrates the pores and upon pyrolysis is reduced to carbon residues which fill the pores, thus reducing porosity).

Diffusivity δ (m^2 s^{-1}) is a good candidate for porosity assessment since it is related to mass density:

$$\delta = \frac{K}{\rho C} \tag{8.1}$$

where K is the thermal conductivity (W m^{-1} °C^{-1}), ρ the density (kg m^{-3}) and C the specific heat (J m^{-3} °C^{-1}). Since diffusivity can be easily measured in a noncontact fashion through infrared thermography, it is thus an attractive technique to discuss. It will be the subject of this chapter. Finally, notice that TNDE is not the only method for diffusivity measurement; other methods such as photoacoustic microscopy have also been developed (Balageas et al. 1991).

8.2 Classical Thermal Diffusivity Measurement Method

Several methods have been developed to compute thermal diffusivity from standard infrared thermography experimentation (Degiovanni 1986; Welch et al. 1990). Probably the best known technique was derived by Parker et al. (1961): the method consists of heating a sample and observing the temperature evolution of either the front or back face. If the back face is monitored, Parker et al. showed that for a sheet of thickness L, the time $t_{1/2}$ (this is the time for the back surface to rise to half its maximum value) can be expressed (assuming a one-dimensional heat flow within a semi-infinite medium, an acceptable hypothesis if experimental conditions can reproduce such a scheme) by

$$t_{1/2} = \frac{1.38\,L^2}{\pi^2\delta} \tag{8.2}$$

Reynolds and Wells (1984) and Hobbs et al. (1991) computed $t_{1/2}$ for a wide range of materials. They concluded that for instance, in the case of a copper plate of 1 cm thickness, $t_{1/2}$ is only 0.13 s while it is 33 s for a 1 cm thickness CFRP plate (measurement taken perpendicular to the fibres). In the case of high diffusivity materials such as copper, this imposes a strong requirement on the acquisition apparatus, especially if conventional video equipment is to be used.

The same analysis can be performed if the front surface is observed. In this case, assuming a one-dimensional heat flow within a semi-infinite isotropic medium and after absorption of a Dirac pulse (Eq. (1.2)):

$$\Delta T = \frac{Q}{e(\pi t)^{1/2}} \tag{8.3}$$

where Q is the absorbed energy and e is the thermal effusivity (Eq. (6.10)). A plot of this curve will show a line of slope $-1/2$ on a log–log scale (Fig. 1.2). In the case of a finite thickness sample the linear cooling decay is followed by a horizontal plateau (Fig. 8.1). From this curve the time t^* can be derived. Time t^* corresponds to the point of intersection of the two asymptote lines in the temperature decay plot; it is

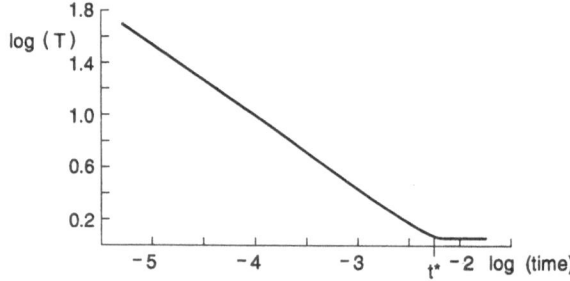

Fig. 8.1. Temperature evolution curve for a finite thickness sample after pulse heating of the surface.

linked to the diffusivity by (Delpech et al. 1990)

$$\delta = \frac{L^2}{\pi t^*} \qquad (8.4)$$

The main problem with diffusivity measurements using front (Eq. (8.4) and back (Eq. (8.2)) surface approaches is to estimate time $t_{1/2}$ or t^*, especially if thermograms are noisy.

8.3 Diffusivity Measurement Method Based on the Laplace Transform

The Laplace transform, well known from electric circuit analysis, found recently some new fields of application in TNDE (see e.g. Houlbert 1991). Delpech et al. (1990) improve significantly the classical thermal diffusivity measurement through the introduction of the Laplace transform. The transform, through the integration it implies, substantially reduces the noise and enables finer diffusivity measurement.

In Laplace space, Eq. (8.3) becomes

$$\bar{T} = \frac{Q \operatorname{ch}(qL)}{Kq \operatorname{sh}(qL)} \qquad (8.5)$$

with the same notation as in Sect. 8.2, with $q = \sqrt{s\delta}$; s is the Laplace variable. A plot of the normalized curve $(s\bar{T})/(s_0 \bar{T}_0)$ as a function of $L\sqrt{s}$ reveals two asymptotes (as in the case of Fig. 8.1 in the temporal domain) from which the intersection point can be found in s^* (s_0 is the limit value when s tends to zero, as for T_0). Diffusivity is then given by

$$\delta = (L\sqrt{s^*})^2 \qquad (8.6)$$

Figure 8.2 illustrates the method. Since the Laplace transform is obtained through integration, a substantial reduction in the noise present in the thermograms is obtained. The reliability of the diffusivity evaluation is thus improved.

We recall that the Laplace transform of the thermogram $T(t)$ is obtained by numerically computing the integral

$$T(s) = \int_0^\infty T(t) e^{-st} \, dt \qquad (8.7)$$

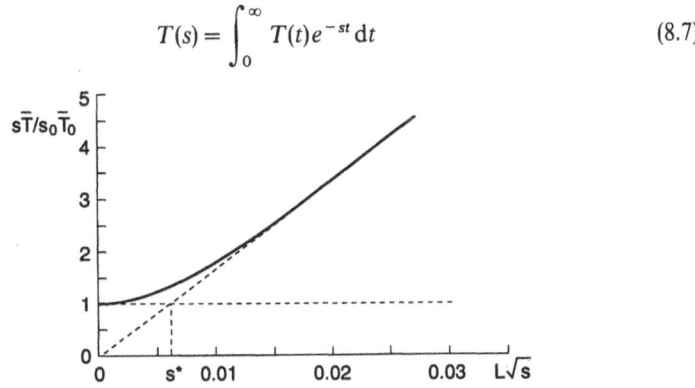

Fig. 8.2. Temperature evolution curve in the Laplace transform domain (see the text).

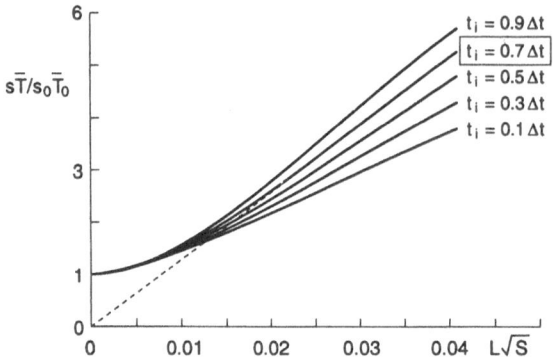

Fig. 8.3. Temperature evolution curve in the Laplace transform domain for different values of the initial time t_i.

As specified in Delpech et al.'s paper, there is a problem in computing Eq. (8.7) since the integral must be evaluated from time zero. This is a problem since the first thermogram acquisition starts at time Δt which corresponds to one acquisition period. They solve the problem by separating Eq. (8.7) into three terms (with t_{max} being the time for end of acquisition):

$$T(s) = \int_0^{t_i} T(t)e^{-st}\,dt + \int_{t_i}^{t_{max}} T(t)e^{-st}\,dt + \int_{t_{max}}^{\infty} T(t)e^{-st}\,dt \qquad (8.8)$$

Second and third terms can be computed without difficulty. The first term can be computed by adjusting the constant p ($t_i = p\Delta t$) so that the asymptote crosses the origin (Fig. 8.3). In fact in this figure, the great sensitivity of the transformed thermogram with p is observed.

In order to determine s^* reliably, that is to determine the intersection point of the two asymptotes, it is necessary to evaluate $s_0 \bar{T}_0$ close to the origin in order to normalize the curve as shown in Fig. 8.2. This can be done by performing a second order fit for values close to the origin (App. C) and extrapolating the value $s_0 \bar{T}_0$.

Reported results on silicon carbide and on magnesium–carbon composites reveal an agreement with the classical method of Sect. 8.2 of about 6%. It is noticed, however, that the Laplace transform method should provide more accurate results since it is less sensitive to noise.

8.4 Diffusivity Measurement Method Based on Phase Measurement

Heath and Winfree (1989) developed an alternative noncontact method for thermal diffusivity measurements based on the phase delay between a thermal wave source (e.g. pulsed Nd-YAG laser) and a spatially offset detection region. For in-plane measurements, temperature evolution is recorded on the sample surface in reflection (same side as the perturbation) using an infrared camera. The measurement set-up is shown in Fig. 8.4. Through-ply measurement is also possible by monitoring the temperature evolution on the back face. Part of the beam needed as timing reference is directed on to the back face with a beam splitter (Fig. 8.5). A typical spatial profile of the point source heating obtained by taking a cross-section of a thermogram is plotted in Fig. 8.6 (carbon–carbon composite material). Figure 8.7 shows a typical temperature evolution of a point spatially offset from the source.

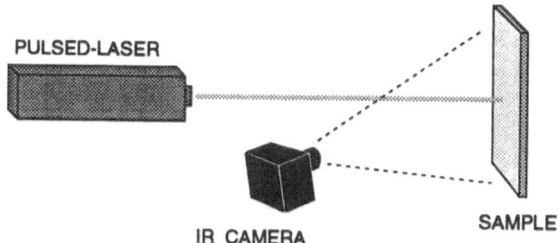

Fig. 8.4. Experimental apparatus for in-plane diffusivity measurement.

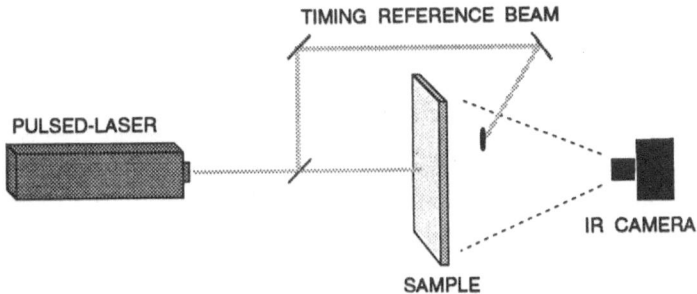

Fig. 8.5. Experimental apparatus for through-ply diffusivity measurement.

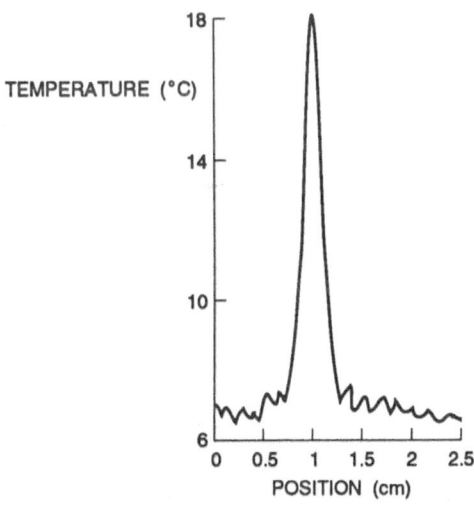

Fig. 8.6. Spatial temperature distribution of the point source (in-plane measurement).

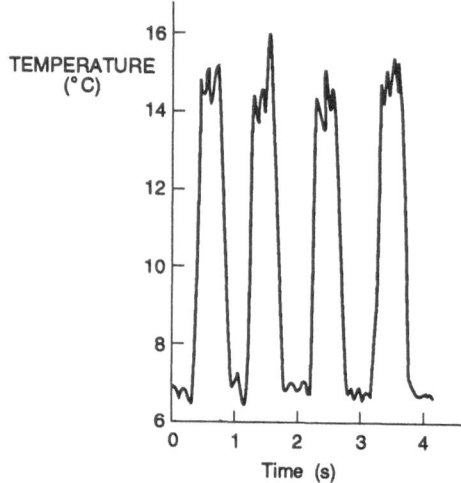

Fig. 8.7. Typical temperature evolution of a point spatially offset from the source (in-plane measurement).

Notice that stimulation performed with an Nd-YAG laser operating at 1.06 μm is advantageous since it is invisible for either a shortwave (3–5 μm) or a longwave (8–12 μm) infrared camera. Reported frequencies of heating pulses are 1 Hz for front face measurement and 0.08 Hz for back face measurement. Low diffusivity in the through-ply measurement is responsible for this low frequency heating. The standard video rate of 60 or 50 Hz limits the application of this method to low to medium diffusivity materials.

If a sample is stimulated with a periodic heat source, Carlslaw and Jaeger (1959) demonstrate that observation of the temperature at a position sufficiently far from

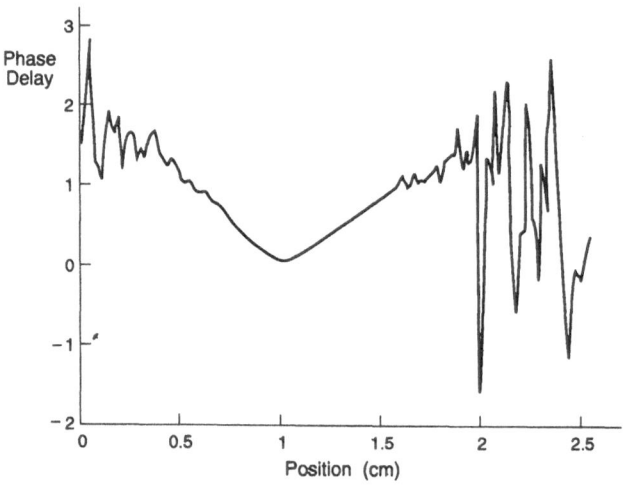

Fig. 8.8. Typical phase delay plotted as a function of the heating source distance (in plane-measurement).

the source approaches a linear function. The slope m of this line is then proportional to the thermal diffusivity of the material (Heath and Winfree 1989, Eq. 1):

$$\delta = \frac{\pi f}{m^2} \qquad (8.9)$$

where f is the frequency of the periodic heat source.

For instance, a typical plot of the phase delay as a function of the heating source distance is shown in Fig. 8.8. In this figure, it is seen that the phase function becomes roughly linear for a certain distance to the source. At greater distances, the SNR becomes smaller and no measurement is possible. From this figure, the slope m of the line can be computed through standard fitting procedures and the thermal diffusivity is obtained using Eq. (8.9).

Diffusivity values obtained using this phase-delay method are reported to be in good agreement with values obtained with other methods (such as those of Sects 8.2 and 8.3).

Conclusion. In this chapter, we have reviewed some infrared imaging methods of interest for thermal diffusivity measurements.

Chapter 9
Thermal Tomography[1]

9.1 General Considerations

Computed X-ray tomography (CT) is now a well established method used to reconstruct the inner structures of components. It has the advantage over conventional X-ray radiographs of easier interpretation of images, since the component under inspection can be "sliced" and observed to any depth. With simple X-ray images, all the inner structure information is compressed. In CT imagery a set of X ray radiographs is recorded along different projections around the component and a special algorithm is then used to reconstruct the unique distribution of attenuation coefficients inside the component (Kassab and Hsieh 1987; Jacoby and Lingenfelter 1989; Garvie and Sorell 1990). This map of attenuation coefficients corresponds to the inner structure of the part expressed as the ability of the matter to attenuate the rays.

The tomography principle cannot be applied directly to the heat transfer process which occurs not along a straight direction but according to a diffusion propagation scheme (Eq. 2.12, Fig. 2.7). Nevertheless, the idea to "slice" a sample into different layers corresponding to the distribution of thermal properties at specific depth layers is interesting. This idea was proposed by Vavilov in 1986; he has since explored its possibilities in collaboration with several research groups worldwide (Vavilov 1990; Vavilov et al. 1990, 1991; Favro et al. 1991; Vavilov 1992; Vavilov et al. 1992).

Instead of being based on angular projections as for CT (Tam 1983), thermal tomography (TT) is based on the surface temperature evolution of the component under inspection following the initial thermal perturbation. In analogy with CT, time increments may be associated with angular projections. In fact TT is just a different method of processing the thermogram sequence and presenting the data: component inspection proceeds in reflection as reviewed in Chap. 1 with experimental methods as studied in Chap. 4.

---

[1]Written with the collaboration of Prof. Vladimir Vavilov of the Tomsk Polytechnic Institute, CIS (former Soviet Union).

9.2 Method

The principles of TT can be understood with respect to Fig. 9.1. In this figure, thermal contrasts under sample surface are drawn (observation is in reflection following a brief initial thermal perturbation). Recalling Eq. (1.1) and concepts of Chaps 2 and 6, it is noted that the occurrence of the time of maximum thermal contrast t_{c_max} is proportional to the square of the depth (at least in homogeneous materials), consequently deeper defects will experience longer t_{c_max}. We also remember how thermal contrast $C(i, j, t)$ is computed at time t for a given pixel (i, j) from temperature image T, from Eq. (6.28):

$$C(i, j, t) = \frac{T(i, j, t) - T(i, j, t=0)}{T_{soa}(t) - T_{soa}(t=0)} \qquad (9.1)$$

where T_{soa} corresponds to the surface temperature over a sound area present in the image. The image before heating (at $t=0$) is also subtracted to remove spurious thermal reflections (see Sect. 6.3.3 for more details concerning thermal contrast computation).

Employing the classical inspection scheme of surface flash heating with observation in reflection, it is possible to measure the temperature decay over the component surface. From such a sequence of infrared images it is possible to compute, for every pixel in the field of view, the time when, for a given pixel, the thermal contrast (Eq. (9.1)) is a maximum. This time is called t_{c_max} and the distribution of all the t_{c_max} values for all the pixels in the image is called a *timegram* TGM_{c_max}. Since a timegram has the same dimensions as an infrared image, it can be displayed as an image. Figure 9.2 illustrates the method of computing timegram image TGM_{c_max}: For $\forall i, i = 0, 1, \ldots, (\text{Maxrow-1})$, for $\forall j, j = 0, 1, (\text{Maxcol-1})$:

$$TGM_{c\_max}(i, j) = t_{c\_max}(i, j) \qquad (9.2)$$

From these definitions, it is understood that a timegram is in fact an "image" of the time distribution of a given parameter (e.g. maximum thermal contrast) extracted over the infrared image sequence recorded during the TNDE inspection of a component. Assuming uniform heating of the component surface, areas of the specimen having uniform thermal properties (such as thermal resistance) will cause the parameter of interest (e.g. maximum thermal contrast) to occur in the same time window

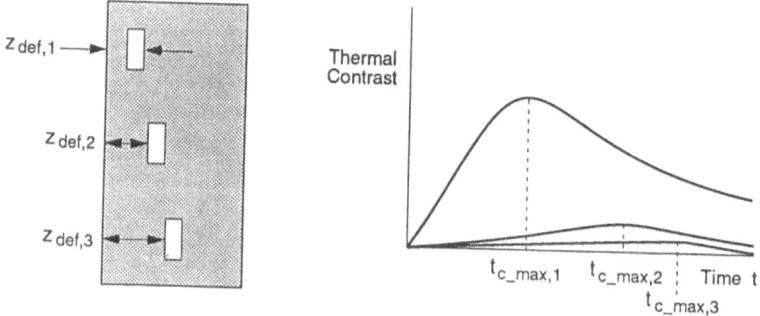

Fig. 9.1. Illustration of occurrence of time of maximum contrast t_{c_max} as a function of subsurface flaw depth.

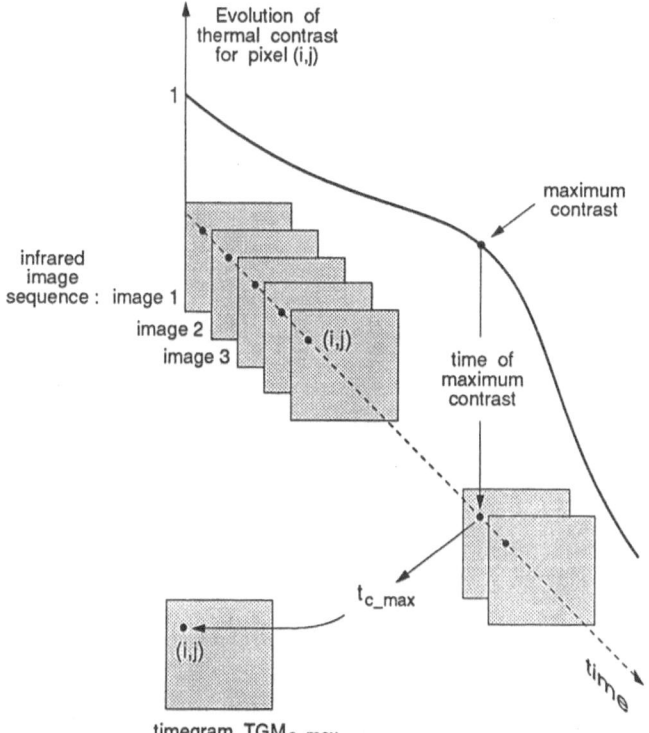

Fig. 9.2. Thermal tomography principle: computation of timegram TGM_{c_max} (see the text).

in the timegram. On the contrary, subsurface flaws having different thermal properties will experience different values of the parameter of interest and consequently, will exhibit different time values in the timegram. This makes defect detection possible.

More interestingly, since the timegram TGM_{c_max} is the time distribution of the occurrence of the maximum thermal contrast computed for every pixel in the field of view and since time of maximum contrast t_{c_max} is proportional to the square of the defect depth, time information in such a timegram is also indicative of the depth of subsurface artefacts present in the component, if any. In order to discover the depths of these artefacts, it is thus necessary to *slice* the timegram, that is to extract from the timegram the values of time of maximum thermal contrast in a given time window $[t_1, t_2]$. A *timegram slice* Slice_TGM_{c_max, t_1, t_2} is obtained by thresholding timegram TGM_{c_max} in a fixed gate time $[t_1, t_2]$:

For $\forall i, i = 0, 1, \ldots, (\text{Maxrow-1})$; for $\forall j, j = 0, 1, (\text{Maxcol-1})$:

$$
\begin{aligned}
\text{Slice\_TGM}_{c\_max, t_1, t_2}(i, j) &= 0 && \text{if } t_{c\_max}(i, j) < t_1 \\
\text{Slice\_TGM}_{c\_max, t_1, t_2}(i, j) &= t_{c\_max}(i, j) && \text{if } t_1 < t_{c\_max}(i, j) < t_2 \quad\quad (9.3)\\
\text{Slice\_TGM}_{c\_max, t_1, t_2}(i, j) &= 0 && \text{if } t_{c\_max}(i, j) > t_2
\end{aligned}
$$

Thus Slice_TGM_{c_max, t_1, t_2} corresponds to a particular layer below sample surface displaying the distribution of thermal properties at the corresponding depth $[z_1 \Leftrightarrow t_1$ to $z_2 \Leftrightarrow t_2]$. Such slice is called a *thermal tomogram* by analogy with the CT tomogram and also because, from the timegram image, the inspected component can be sliced into multiple thermal tomograms corresponding to different depth layers under the surface.

It is important to note that TT cannot "see" a subsurface flaw behind another, since the first flaw prevents the thermal front reaching the second one. However, since the first flaw is identified, it already provides some information about the integrity of the component; consequently this limitation may not necessarily be dramatic. Moreover, TT technique can only be deployed in reflection (one-side inspection), since in transmission, the time information about defect depth is lost (roughly speaking, the time of arrival of the thermal front on the back surface is the same whatever the defect depth).

The possibility to isolate a specific depth layer under surface sample inspected with TNDE is the main attraction of this technique which requires no special apparatus to be employed. In fact, this is only a different way to process infrared images. In fact, from the thermogram sequence recorded over a given inspected component, it is possible to compute different types of timegrams TGM depending on the parameter of interest, either the time of maximum contrast t_{c_max}, the time of half-rise contrast $t_{c\_1/2max}$, the time of half-decay contrast $t_{c\_max1/2}$ or the time at 3 dB of maximum contrast t_{c_3dBmax}

For $\forall i, i = 0,1,\ldots,(Maxrow-1)$; for $\forall j, j = 0,1, (Maxcol-1)$:

$$TGM_{c\_max}(i,j) = t_{c\_max}(i,j) \tag{9.4}$$

$$TGM_{c\_1/2max}(i,j) = t_{c\_1/2max}(i,j) \tag{9.5}$$

$$TGM_{c\_max1/2}(i,j) = t_{c\_max1/2}(i,j) \tag{9.6}$$

$$TGM_{c\_3dBmax}(i,j) = t_{c\_3dBmax}(i,j) \tag{9.7}$$

where $t_{c\_max}(i,j)$, $t_{c\_1/2max}(i,j)$, $t_{c\_max1/2}(i,j)$ are extracted from the thermal contrast evolution curve for pixel (i,j) (Fig. 6.10). As stated in Chap. 6, evaluation of parameters $t_{c\_max1/2}$ and $t_{c\_1/2max}(i,j)$ is often more reliable than evaluation of t_{c_max} because of the slow variation of thermal contrast near its maximum value.

An important issue in TT is to determine the depth resolution which can be achieved, that is in *how many layers a given sample can be sliced into* for a given TNDE experiment. Vavilov et al. (1992) indicated that such depth resolution is limited both by the rate at which infrared images are recorded and also by temperature resolution of the imager (limited by noise). In turn, noise limitation affects depth resolution in two ways: detector noise and surface noise. From these considerations, depth resolution Δz can be estimated by (Vavilov et al. 1992, Eqs 5 and 6):

$$\Delta z = \frac{C_{noise}}{\dfrac{\partial t_{c\_k}}{\partial z} \dfrac{\partial C}{\partial t}} \tag{9.8}$$

where C_{noise} is the thermal contrast noise, $\partial t_{c\_k}/\partial z$ is the variation of the parameter of interest with depth (k stands for the parameter of interest either: time of maximum contrast time of half-rise contrast,...) and $\partial C/\partial t$ is the first derivative of thermal contrast curve. From Eq. (9.8), it is seen that recourse to parameter $t_{c\_1/2max}$ or t_{c_3dBmax} for which $\partial C/\partial t$ is greater than for t_{c_max} is more advantageous since more layers can be resolved (smaller Δz). As an example, in carbon epoxy it can be demonstrated that parameter t_{c_3dBmax} allows resolution of more than twice the depth of layers than parameter t_{c_max}. Finally, depth resolution of TT can also be enhanced by increasing the input energy power needed for component thermàl stimulation since in this case C_{noise} will be reduced. Of course we should always be careful to avoid damaging the sample through surface overheating if we still want to talk about a nondestructive evaluation method!

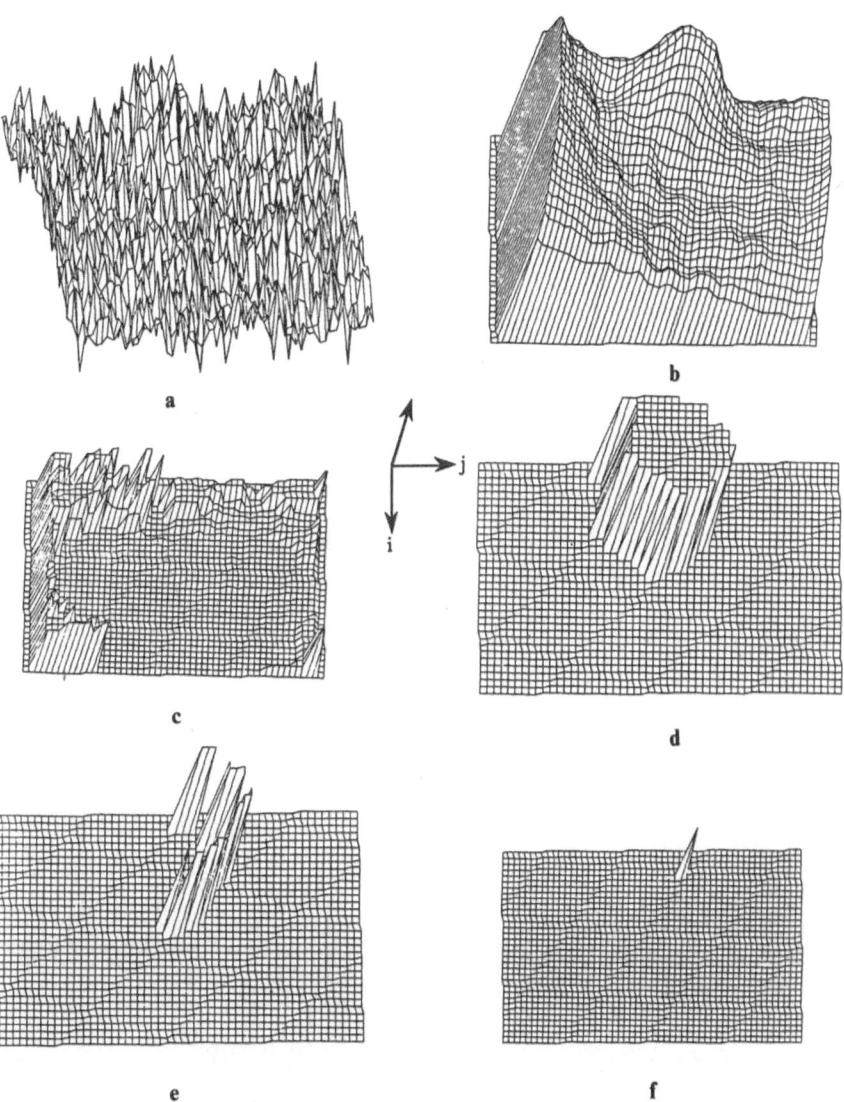

Fig. 9.3. Thermal tomography of Teflon™ insert in carbon epoxy specimen using the t_{c_max} parameter: **a** raw image, **b** smoothed image, **c** timegram TGM_{c_max}, **d** tomogram of the layer 0.8–1.5 mm, **e** tomogram of the layer 1.4–1.8 mm, **f** tomogram of the layer 1.8–2 mm. Specimen: 4.25 mm-thick, 28-layer black-painted CFRP panel with a 10 mm diameter Teflon™ implant inserted at the eighth layer (1.2 mm beneath the front surface). Arbitrary amplitude units.

9.3 Some Results

Some results of the TT technique reviewed in Sect. 9.2 are shown in Fig. 9.3 for the case of a Teflon™ insert in a carbon epoxy specimen. A raw image recorded at the time the contrast is a maximum over the defect ($t_{c\_max} = 4.35$ s) is shown in Fig. 9.3a. The smoothed maximum contrast image reveals a temperature contrast over the defect of about 6% (Fig. 9.3b). Noise reduction techniques studied in Sect. 3.2.2 can be used profitably at this stage. Timegram TMG_{c_max} synthesized from 60 images taken during the specimen's cooling is shown in Fig. 9.3c: in this image, it is clearly

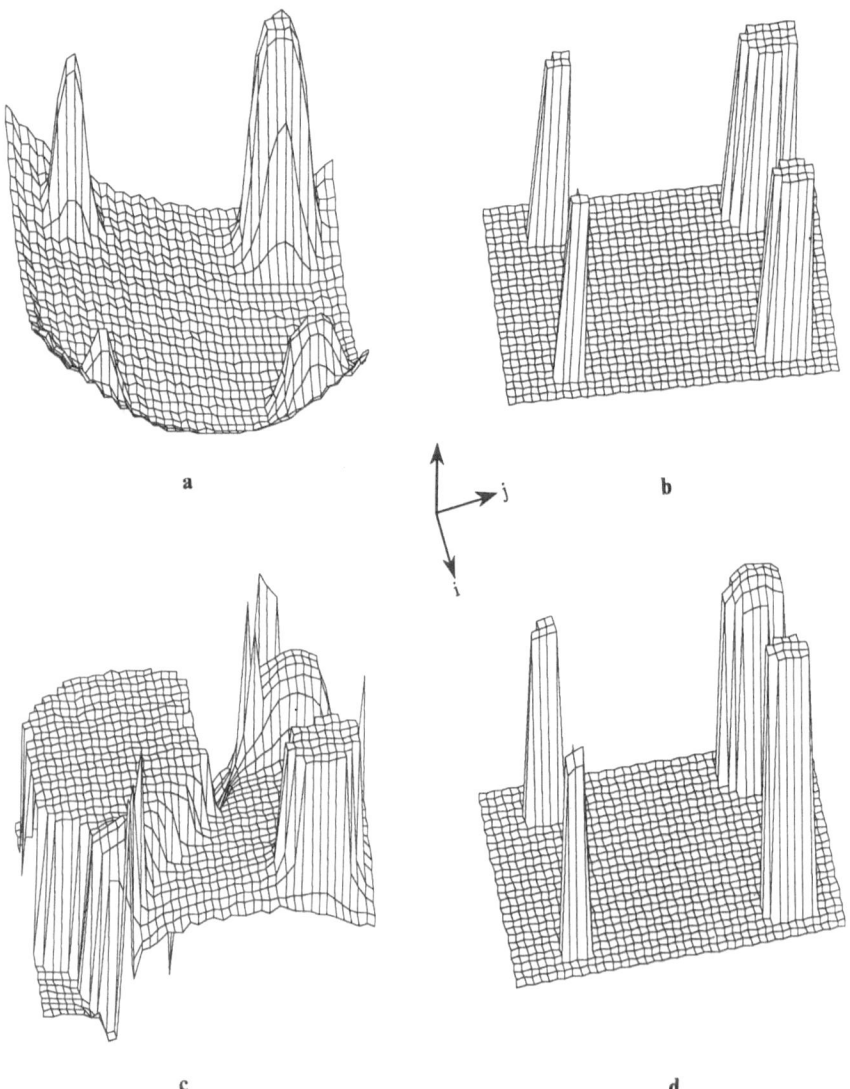

a   b

c   d

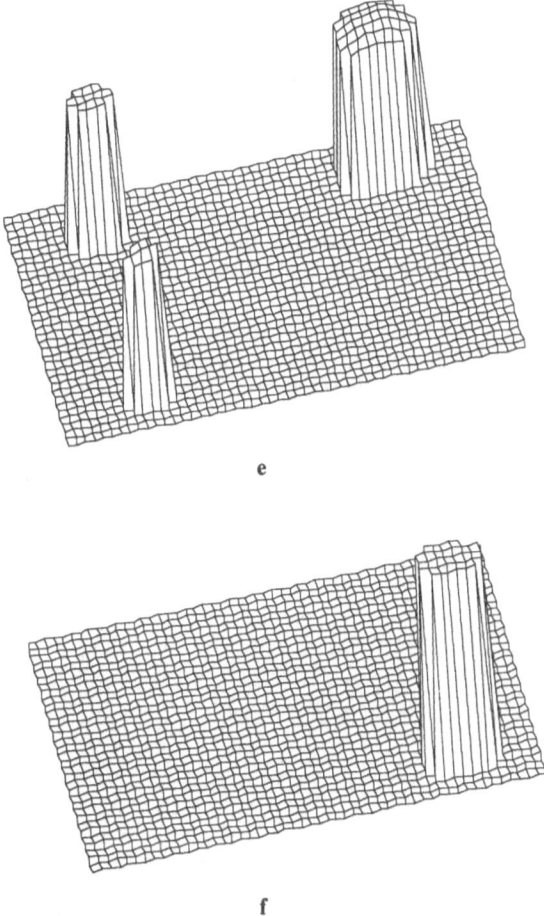

e

f

Fig. 9.4. Thermal tomography of Plexiglass™ plate with four holes drilled from the back surface at different depths: **a** summation of all the raw images (no correction for radiometric distortion); **b** binary image following defect detection; **c** timegram TGM_{c_max}; **d** product of **b** and **c**; **e** tomogram for depths smaller than 2 mm; **f** tomogram for depths greater than 2 mm (only deepest (2.5 mm) defect is visible). Arbitrary amplitude units.

seen that edge irregularities produce higher values of t_{c_max} than the defect area, which is scarcely noticed at this particular scale. Tomograms are shown in Fig. 9.3d–f. Separation of the layer 0.8–1.5 mm which presumably contains the defect shows the Teflon™ insert very distinctly (Fig. 9.3d). The tomogram of a layer 1.4–1.8 mm (Fig. 9.3e) reveals some deeper disturbances around the defect, although strictly speaking this could not be deeper, as stated previously. It is probably just weaker inner irregularities in "t_{c_max}" terms. Finally, the layer 1.8–2 mm (Fig. 9.3f) shows a single signal which cannot be identified; nevertheless this tomogram clearly shows the lack of major defects inside this layer.

An interesting addition to TT is first to include a defect detection step which allows removal of unwanted structures from the tomograms. Figure 9.4 illustrates the method for the case of TNDE of a Plexiglass™ plate in which subsurface flat-bottom holes have been drilled at different depths from the back surface. The evaluation proceeds in two steps: defect detection and tomogram computation. For defect detection, the algorithm of Sect. 4.4.3 is used: Fig. 9.4a is the summation image of all raw infrared images (Eq. (4.3)) (notice uncorrected radiometric effects on image edges), while Fig. 9.4b is the image of Fig. 9.4a after segmentation (with techniques of Sect. 4.4.4). Figure 9.4c is the timegram TGM_{c_max} (Eq. 9.2). Figure 9.4d is the product of Fig. 9.4b and 9.4c which allows suppression of unwanted regions of the timegram. Finally Fig. 9.4e is the tomogram for depth layer 0–2 mm (0–50 s) and Fig. 9.4f is the tomogram for depth layer greater than 2 mm (> 50 s). In Fig. 9.4f the deeper 2.5 mm defect is separated from the three other shallower subsurface holes ($z_{def1,2,3} < 2$ mm).

Conclusion. The TT technique provides another way to look at thermal information in terms of time rather than amplitude as in standard TNDE image processing. It has some advantages concerning interpretation of results, especially since detected structures appear directly in terms of depth (after conversion from the time domain). Among the drawbacks of the technique, we can mention the amount of computation needed since all images must be processed. Other limitations are as follows: strong reduction of spatial resolution with depth; detection limited to subsurface artefacts having different thermal properties with respect to the bulk of material. Note, however, that these limitations are, of course, shared by the standard TNDE procedure as well.

Chapter 10

Thermal NDE of Nonplanar Surfaces

10.1 General Considerations

Thermal inspection of nonplanar surfaces is a subject of interest since many components susceptible to be inspected by TNDE are of complex geometries. Such components can be divided roughly into two categories: approximately planar; and nonplanar. Obviously, the dividing line between these two categories is not clear (Duda et al. 1979). An arbitrary criterion is for instance that for all the points on the inspected surface, the normal to the surface is contained in an angle

$$\pm 3\,\text{dB of }90° \qquad (10.1)$$

This restricts the cosine of the normal to the [0.75, 1] interval (this hypothesis, as we will see below, allows some practical assumptions). In fact the distinction between planar and nonplanar samples is relevant since, for approximately planar components, TNDE methods studied in the other chapters of this book can be applied without much difficulty.

Another important concern is the size of the objects to be inspected: nonplanar specimens of large size can be asssumed planar in limited fields of view. In these circumstances it is possible to move the thermographic inspection head (flash lamps and infrared camera) over the specimen and orientate this apparatus to respect the criterion of Eq. (10.1). It is, for instance, possible to mount the apparatus on a robot arm, thus facilitating the moving process as Délouard et al. (1989) and Tretout et al. (1991) suggest. In this case, an accurate way of moving the apparatus over the specimen surface must be found, which maintains both proper orientation and distance from the surface. For example:

Ultrasonic range finder (low cost, but resolution restricted to a few millimetres over small distances of some tens of centimetres)

Three-dimensional (3D) camera (high cost, but provides a very accurate map of the surrounding environment, see for instance Maldague et al. (1986a) Rioux et al. (1987); Blais et al. (1988); Poussart and Laurendeau (1988); Soucy and Laurendeau (1990) for more details on 3D perception)

Three-dimensional analysis derived from conventional black-and-white 2D camera using the shape from shading concepts (see for instance Bruckstein 1988)

Implementing 3D perception using the available infrared equipment (this can be achieved to some extent as we will discuss in Sect. 10.2; see also Nouah et al. 1993).

In this chapter we will briefly discuss the issue of TNDE inspection of nonplanar regions of components. This inspection must be handled properly for the interpretation of thermograms to be correct. Consider for instance the possible misinterpretation which can occur in the case of uneven heating due to local surface curvature. To a certain extent such local hot spots can be falsely assumed to be subsurface flaws.

This topic is rather new in TNDE; consequently, we will limit ourselves to a general introduction: guidelines given below will surely evolve in the future. Nevertheless, we will try to provide some insight into the topic.

10.2 Principle of Surface Curvature Correction

We will assume two kinds of radiative thermal stimulation: dispersed rays or parallel rays. A dispersed source of energy can be assimilated to a point source (for example a light bulb) for which the energy deposited is a function of the square of the distance between the source and the stimulated surface (put your hand close to and then far from a light bulb and notice the difference in temperature on your hand). For a point source the density of power P_d in $W\,m^{-2}$ deposited can be expressed as

$$P_d = \frac{P}{4\pi R^2} \tag{10.2}$$

in which P is the power of the source in watts and R is the source–surface distance.

If a parallel ray source is used, then P_d is a constant with R, at least over restricted distances. An example of a parallel ray source used for TNDE is a cinematographic spotlight with back parabolic reflector (Fig. 3.6).

With respect to Fig. 10.1, we can write for a surface patch dA_{xy} tilted by θ degrees with respect to incident rays and located at distance R from the point source of power P, that power $P_{A_{xy}}$ received is given by

$$P_{A_{xy}} = P_d dA_{xy} \cos\theta \tag{10.3}$$

and the associated temperature rise ΔT_p is then given by (Thomas 1980, Eq. 3.40)

$$\Delta T_p = \frac{P_{A_{xy}}\Delta t}{\rho C_p dz\, dA_{xy}} \tag{10.4}$$

where

Δt thermal pulse length, Δt (s)
$P_{A_{xy}}$ power received on surface patch dA_{xy} (W)
ρ material density ($kg\,m^{-3}$)
C_p specific heat ($J\,kg^{-1}\,°C^{-1}$)

Combining Eqs (10.3) and (10.4), we obtain

$$\Delta T_p = \frac{P_d \cos\theta\,\Delta t}{\rho C_p dz} \tag{10.5}$$

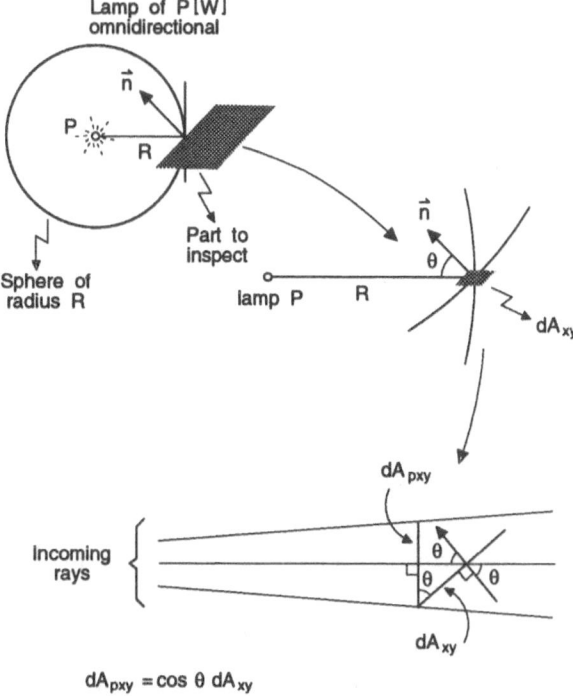

$$dA_{pxy} = \cos \theta \; dA_{xy}$$

Fig. 10.1. Geometry for energy deposition computation.

where dz is the depth of penetration of the thermal front under the surface. It may be evaluated with the following assumption (for homogeneous materials, Eq. (1.1)):

$$dz \approx \sqrt{\delta \, \Delta t} \qquad (10.6)$$

where δ is the thermal diffusivity (m^2 s^{-1}). Table 1.1 gives thermophysical parameters for common materials. In Table 10.1 we computed some values of depth of penetration for two materials having very different thermal behaviour (alumininum with high thermal diffusivity and Plexiglass™ with low thermal diffusivity).

Table 10.1. Estimation of depth of penetration of the heat front under the surface as a function of time for various materials

Length of thermal pulse Δt (s)	Depth of penetration dz	
	Aluminium (mm)	Plexiglass™ (mm)
0.01	1	0.04
0.1	3	0.1
0.5	7	0.3
1	10	0.4
2	14	0.6
100	100	4

As stipulated by Eq. (10.5), the temperature rise is directly proportional to the power source P_d, thermal pulse length Δt and surface orientation $\cos \theta$. Since the measured parameter ΔT_p should be as large as possible for greater signal-to-noise ratio, not considering parameter $\cos \theta$ for now, from Eq. (10.5) we see that two choices are possible: increase P_d or Δt; Δt cannot be too large since a large value of Δt leads to a large depth of penetration (Table 10.1) and a large depth of penetration implies the thermal front may have already interacted with subsurface defect(s) causing surface temperature variations to be picked up by the camera which do not depend only on the object shape or the thermal stimulation apparatus. Since, as a first step, we only want to correct images for shape curvature and thermal stimulation, effects of possible subsurface defects are not of interest and should be limited as much as possible. Consequently a small depth of penetration dz is desired requiring small Δt and large P_d.

From Eq. (10.5), many different situations can be considered. For instance, in the case of the repetitive inspection of components of known geometry, the $\cos \theta$ parameter, once the component is located in the field of view, becomes known (e.g. from CAD/CAM files). It is then interesting to use a dispersed source of energy for component thermal stimulation. For such a source, the energy deposited is a function of the distance (Eq. 10.2). Since, during the heating pulse, the temperature rise over the surface is also a function of the distance, it is then possible to obtain the camera–surface distance R from the measurement of the surface temperature rise, taking into account the perspective effect.

As the next step, the emitted energy from the component surface is also dependent on R and $\cos \theta$ with relationships similar to Eqs (10.2), (10.3) and (10.4). Knowledge of these parameters allows correction of the incoming images for an uneven rise in temperature. Consequently, defect detection and characterization (through thermal contrast computation) would be enhanced, thus improving TNDE performance.

If object geometry is unknown, both dispersed and nondispersed sources of energy can be used. Using the nondispersed source of energy (no dependence on R to a certain extent), the $\cos \theta$ factor can be estimated from the surface temperature rise due to the heat pulse. The dispersed source of energy can then be used for the analysis described in the previous paragraph.

From this very informal discussion, we come to the conclusion that in addition to the classical defect evaluation aspect, it is possible to extract some gross 3D information from the TNDE procedure. In turn this information can help in improving TNDE system capabilities.

Figure 10.2 illustrates such an approach; the experimental configuration is presented in Fig. 10.2a. Two images are presented in Fig. 10.2b: the left one is the temperature image recorded over a Plexiglass™ plate tilted in front of the infrared camera. A 1000 W dispersed source of energy positioned on top of the infrared camera is used to flash-heat the surface. On this thermogram two temperature trends are clearly visible: uneven temperature from left to right and uneven temperature from top to bottom. Since the plate is tilted, the right corner is further than the left corner with respect to the source and consequently receives less energy, thus explaining the temperature distribution observed. The right image is a 3D range image of this scene computed using the principles described in this section. On this image the top-to-bottom trend disappears; only the left-to-right trend is noticeable, which is due to the plate orientation with respect to the infrared camera. A limited increase in temperature is obtained on the tilted plane due to the dispersed nature of the stimulation source used,

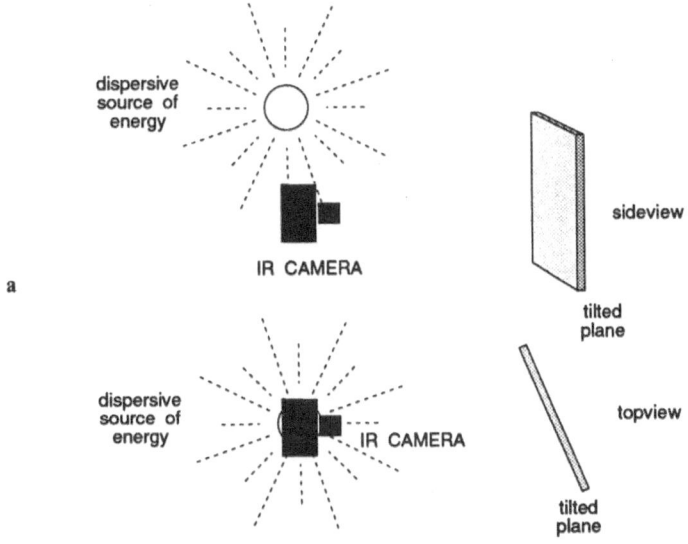

a

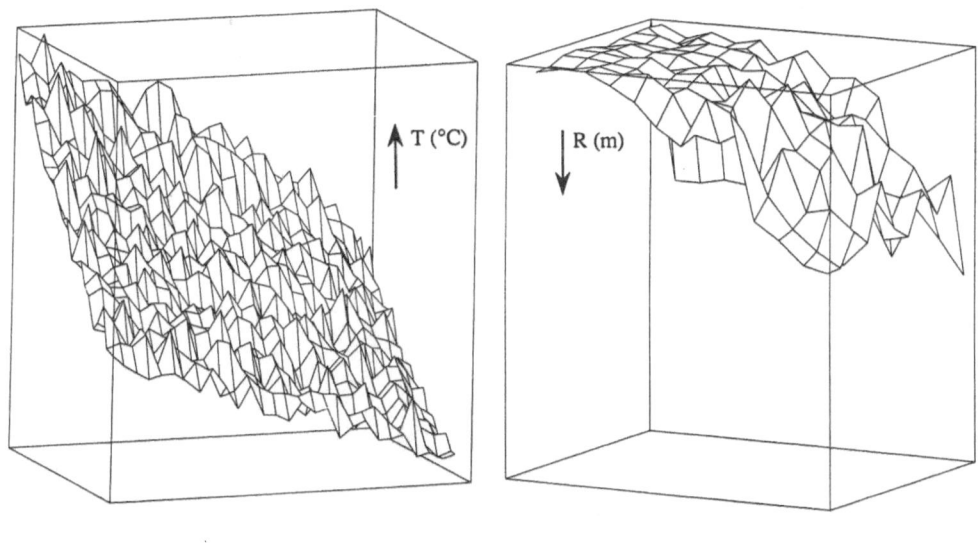

b

Fig. 10.2. Thermal NDE for nonplanar surfaces: **a** experimental configuration; **b** experimental results obtained on a tilted Plexiglass™ plate: left image is the temperature image while right image is the computed three-dimensional range image of this scene (arbitrary units).

consequently the distance computation rapidly degrades, as seen on the right of the range image.

Conclusion. This chapter has introduced the topic of TNDE inspection of nonplanar surfaces. This field is not yet developed and is still the subject of research activities. In this respect, it is important to note that the discussion was intentionally simplified and somewhat idealistic, but is still able to give some insight into this matter.

Chapter 11

Applications of Infrared Thermography to High Temperatures

In this last chapter we will briefly discuss high temperature applications of infrared thermography. Strictly speaking, this is not a prime field of application for what we have called *active* TNDE. However, since infrared equipment can be employed in high temperature contexts, it was found interesting to include a short discussion on this subject in this book (for instance both AGEMA model 880 and Inframetrics model 760 radiometers are capable of measurement up to 1500 °C).

In this chapter, we do not pretend to cover this topic in any extensive manner; we will restrict ourselves to two case studies. Of course, many other applications could be envisaged.

11.1 Detection of Rolled-in Scale on Steel Sheets

Scale patch formation on steel sheets occurs during the reheating step prior to hot rolling (at about 1300 °C in the reheating furnace). These scale patches can reach up to 1 mm thickness and are formed essentially from FeO, FeO_4 and Fe_2O_3. Chemical, mechanical and hydraulic techniques such as acid pickling, steel surface sweeping with chains and high pressure water spraying are used in the steel industry to prevent oxide formation and remove scale patches. However, if adherence of patches on steel is good, they can stay on the surface and thus be rolled in the steel sheet during hot rolling. Figure 11.1 depicts the problem. Such steel sheets have less aesthetic appeal which reduces their commercial value. Even if painted, rolled-in scale patches remain visible on steel sheets. Figure 11.2 shows pictures taken in the visible spectrum of rolled-in scale (5 mm × 30 mm): the elongated shape is due to the rolling process which tends to spread the scale material along the rolling direction and scale crusts formed on the steel sheet surface after furnace heating.

An inspection system which can recognize the presence of scale before and after rolling would be very interesting for process and quality control. Often such inspection is performed visually by an operator after the fact thus causing expensive production of wasted material.

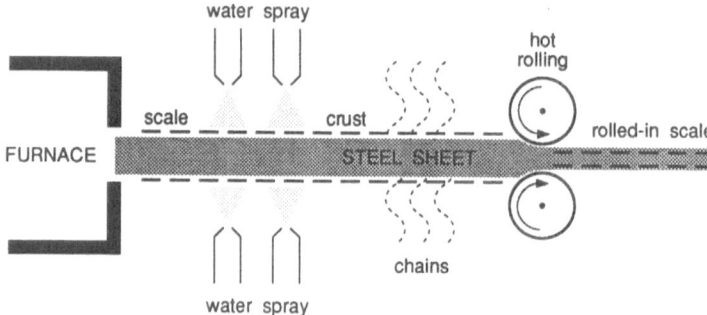

Fig. 11.1. Formation process of scale crusts on steel sheets during hot-rolling operations.

a   b

Fig. 11.2. Pictures taken in the visible spectrum of **a** rolled-in scale (5 mm × 30 mm): the elongated shape is due to the rolling process which tends to spread the scale material along the rolling direction; **b** scale crusts present on the surface prior to rolling.

Obviously, considering the hostile environment of the hot rolling process (high temperature, heavy water vapours, large vertical excursions of the sheets), a non-contact technique is required.

The thermographic approach is a good candidate for this application, taking advantage of the naturally high temperature of the steel sheets during hot rolling.

If the scale patches are not perfectly adherent to the steel surface, the temperature distribution recorded over the sheet allows detection of residual scale patches which are at a lower temperature than the metal, because of the larger thermal resistance

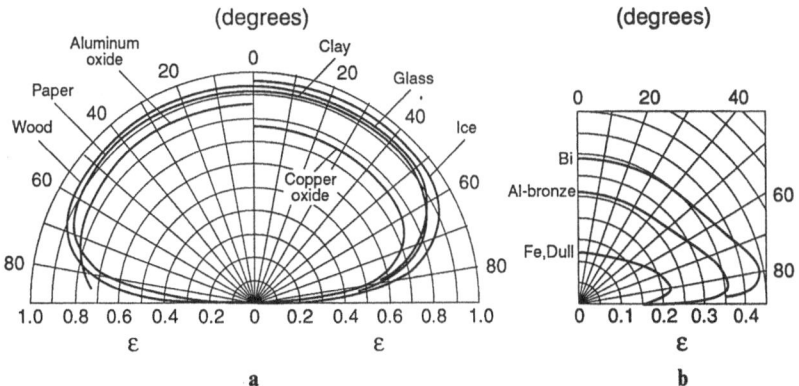

Fig. 11.4. Polar diagrams of the infrared emissivity of a number of **a** nonmetallic and **b** metallic materials.

of the scale–steel interface (Cielo 1984). However, high scale emissivity with respect to steel can compensate for this temperature difference. It is then necessary to rely on active thermal stimulation by cooling the surface using a gas jet, taking advantage of the low thermal conductivity of the scale whose temperature will be reduced significantly while bulk steel will stay at a higher temperature (this is the *cool wave* approach described in Sect. 4.3). As a general rule, thermal resistance effects are more important at the descaling station where scale patches are not yet rolled-in within the metal. On the contrary, after rolling, thermal resistance is reduced and consequently emissivity effects are prominent. These principles can be analysed experimentally.

In order to demonstrate the inspection method, steel sheets were uniformly heated by immersion in a warm water tank (water at about 5 °C above room temperature). Figure 11.3 (see colour section) shows the comparison between a video image of rolled-in scale crust (visible image, diffuse illumination) and an infrared image. The infrared image shows more clearly the high emissivity defect and is less sensitive to surface discoloration patterns which are clearly noticeable in Fig. 11.3a (they seem to be due to FeOOH deposits formed by the repetitive contact of the steel with warm water during the experiments). At room temperature, scale defect visibility is much lower because emissivity differences are compensated by opposite variations in the surface reflectance from the ambient temperature background.

When warm steel sheets are inspected with an emissivity variation criterion related to the presence of rolled-in scale effects, the observation angle is an important factor to take into consideration. Figure 11.4 displays polar emissivity diagrams for common industrial materials. In this figure, we notice that emissivity is fairly constant for nonmetallic material for a wide variation in the observation angle from normal incidence. For smooth metallic surfaces, emissivity tends to be lower at normal incidence than at grazing incidence. This suggests that rolled-in scale defect visibility will be higher at normal incidence when emissivity of the smooth steel surface is low with respect to scale emissivity, which is also uniform. This result is confirmed in Fig. 11.5 (see colour section) which compares emissivity of steel sheet and rolled-in scale defects for different orientations with respect to the infrared camera.

Figure 11.6 shows a comparative temperature evolution curve for two adjacent zones over a steel sheet: one above a rolled-in scale defect and the other over the smooth steel surface. These temperatures were recorded while an air jet was cooling

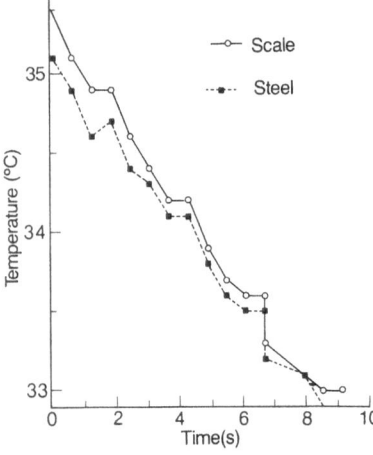

Fig. 11.6. Comparative temperature evolution of two points, one over a rolled-in scale patch and one over bare steel without scale while steel sheet is subjected to a cool air stream.

the surface. It is noticed that the apparent temperature of scale is always superior to the apparent steel temperature; however, the temperature differential between the two curves tends to reduce as sheet temperature reduces. These temperature differences are partly due to emissivity differences as indicated above and confirm also the low thermal resistance at the scale–steel interface. Greater temperature differences are present if scale patches are partly loose and the surface is observed under forced cooling conditions; Fig. 11.7 (see colour section) illustrates this. A steel sheet covered with scale patches formed during the passage in the furnace is observed with the infrared camera while air jets cool the surface. The temperature of the steel sheet is slightly above room temperature and the steel surface was partly descaled so that free scale steel areas are present, as seen in Fig. 11.2b. In Fig. 11.7, areas of low temperature correspond to loose scale patches which have a greater thermal resistance at the scale–steel interface. Areas of higher temperature correspond to bare steel whose thermal inertia prevents fast cooling.

This cool thermal stimulation approach has many advantages as mentioned previously: low cost of thermal stimulation apparatus; improved security; no spurious thermal reflection from the thermal source. Note that for this application, air is not a good stimulation medium since it activates scale formation at such high rolling temperatures; consequently, an inert gas is preferred.

This case study shows that infrared thermography is an attractive technique for scale detection either before or after rolling. Rugged infrared imaging equipment is now commercially available and can be used in a harsh environment such as that found in the steel industry.

In addition to the scale detection method presented here, it should be pointed out that temperature measurement of the hot-strip during cooling is important in order to maintain a certain grain size before and after the transition from one metallurgical structure (e.g. thick slab) to another (e.g. thin sheet). In fact, Holmsten and Houis (1990) specify the various locations where temperature measurement is critical in the steel industry: at the exit of the reheating furnace, at the rougher and descalers (as specified above), after the last finishing stand and before coiling.

Fig. 11.8. Passive infrared thermography used in designing heat exchanger units for warehouses.

11.2 Thermal Inspection of High Temperature Industrial Structures

As a second case study, we will present an example of passive thermography (cf. Sect. 1.1). We recall this approach does not require a source of thermal perturbation. In fact, the inspected structure is already at a given temperature and the temperature distribution recorded over its surface contains useful information regarding its operating regime or its physical integrity.

In this section, we will see how infrared thermography can be useful to help solve design problems in the case of heat exchangers used to warm warehouses. Figure 11.8 shows such a unit. The unit is made of aluminium–steel elements internally heated by gas burners. The heat is transmitted to the ambient environment by forced air circulation between the elements by means of a fan located at the rear of the unit.

A thermal image of the unit brings a map of the heat distribution within the elements. An example of such a thermogram is shown in Fig. 11.9 (see colour section). Hotter areas are visible in the centre at a position determined by the geometry of the convection coefficient, which depends on hot gas circulation inside the elements. A slight displacement of the maximum temperature distribution towards the right part of the unit is also visible on this thermogram. This is due to the counterclockwise rotation of the propeller fan.

Many tests can be performed with the heat exchanger operating under various regimes and with different orientations of the vent panels (visible as cold structures in Fig. 11.9). The heating rate during the start-up or shut-down period can be analysed as well, in order to specify optimum operating conditions such as burner/fan turn-on cycles for both fast response and reduced thermal-shock fatigue of specific elements.

In general such analysis is a very powerful tool to assist the designer in the conception of new units and to locate areas of extreme thermal concentration which require reinforcement or geometrical modification.

Conclusion. In this chapter we have briefly discussed two examples of active and passive TNDE deployment in high temperature contexts.

Appendix A
Computer Model

This computer modelling program is discussed in Chap. 2.

```
/*****************************************************************/
/*
                          SIMUL.C

Computation of the temperature distribution in a CFRP specimen

          radial conductivity ra_conduct:.05w/KgC
          axial  conductivity ax_conduct:.01w/cmC;
          specific heat: 1000 J/KgC
          mass density:  2000 Kg/m3
          Thermal losses geometric factor:
              Frad:5.67*10e-12w/cm2;
              Fcon:10e-3

    Program adapted in Borland "C" language by Mr. Ahmed NOUAH
    from Engineering Dept, Université Laval, Québec city, Québec
    Canada G1K 7P4.

    Special thanks are          acknowledged to Dr. Paolo CIELO from
    Industrial Materials        Research Institute, Boucherville,
    Canada for his ideas and program examples on heat transfer
    modelling.

    This computer program work  on PC or Sun type of computer.
    Results are displayed on the computer screen. On a PC, the
    output can be redirected in a file using the following
    command:

          simul > filename

          where "simul" is modelling program name
          and "filename" the name of the file to save results

    *****************************************************************/
#include <stdio.h>
#include <math.h>
#include <alloc.h>
#define pi 3.1415
#define Sqr(a) ((a) * (a))
```

```c
#define Sqr4(a)  ((a)*(a)*(a)*(a))
main()
{
/* thermal parameters */

#define rad_nodes 20      /* number of radial nodes */
#define axe_nodes 30      /* number of axial nodes */
#define def_nod_pos 11    /* number of axial defect nodes */

int i,j,l,iter;
double ***temp,**rhd,**cp;
double **qr,**qz,*kint;
double *dslat,*dslong;
double dr,dz,dt,tim,timlc,timl,timc,source_power;
double temp_var,red_loss,conv_loss,ra_conduct,ax_conduct;

/*memory allocation */
kint=(double*)malloc(rad_nodes*sizeof(double));
dslat=(double*)malloc(rad_nodes*sizeof(double));
dslong=(double*)malloc(rad_nodes*sizeof(double));

rhd = (double**)malloc(rad_nodes*sizeof(double*));
cp  = (double**)malloc(rad_nodes*sizeof(double*));
qr  = (double**)malloc(rad_nodes*sizeof(double*));
qz  = (double**)malloc(rad_nodes*sizeof(double*));
temp=(double***)malloc(rad_nodes*sizeof(double**));

for(i=0;i<=rad_nodes;i++){
        temp[i]=(double**)malloc(axe_nodes*sizeof(double*));
        for(j=0;j<=axe_nodes;j++)
        temp[i][j]=(double*)malloc(2*sizeof(double));

        rhd[i] =(double*)malloc(axe_nodes*sizeof(double));
        cp[i] = (double*)malloc(axe_nodes*sizeof(double));
        qr[i] = (double*)malloc(axe_nodes*sizeof(double));
        qz[i] = (double*)malloc(axe_nodes*sizeof(double));
                }

for(i=0;i<rad_nodes;i++){
        kint[i]=0.01;
for(j=0;j<axe_nodes;j++){
        for(l=0;l<2;l++)
                temp[i][j][l]=0.00001;
        rhd[i][j]=2.0;
        cp[i][j]=1.0;
        qr[i][j]=0.00000001;
        qz[i][j]=0.00000001;
        }}


/* Defect parameters */
for(i=0;i<5;i++) kint[i]=0.00024;/* defect is air*/

/* Modelling geometry */
dr=0.2;
dz=0.02;
dt=2.5e-2;

for(i=0;i<rad_nodes;i++){
        dslat[i]=2.0*pi*(i+1)*dr*dz;
        dslong[i]=pi*(Sqr((i+1)*dr)-Sqr((i)*dr));
        }

/* beginning of time iteration loop
   "tim"  is the current time
```

```
        "timc" indicates when a display of results is to be done
*/
tim=0.0;   /* starting time */
timc=0.4; /* first time to display result */
timlc=log10(timc);
for(iter=0;iter<=1200;iter++){                  /*1200 iterations*/
            tim+=dt;
            timl=log10(tim);

/* Thermal perturbation source  */
/* Application of thermal pulse for 2s on the surface (Z=1)      */
source_power=1.0;   /*    power is in W/cm^2 */
if(tim>2.0) source_power=0.0;
for(i=0;i<rad_nodes;i++){
            temp_var=(source_power*dt)/(rhd[i][0]*cp[i][0]*dz);
            temp[i][0][0]+=temp_var;
            temp[i][0][1]=temp[i][0][0];
            }

/* heat propagation through the specimen */
for(i=0;i<rad_nodes-1;i++)
for(j=0;j<axe_nodes-1;j++){
            ra_conduct=0.05;/* thermal conductivity */
            ax_conduct=0.01;
            if(j==def_nod_pos) ax_conduct=kint[i];

            qr[i][j]=ra_conduct*dslat[i]*(temp[i][j][0]-
                                         temp[i+1][j][0])/dr;
            qz[i][j]=ax_conduct*dslong[i]*(temp[i][j][0]-
                                         temp[i][j+1][0])/dz;

            temp[i][j][1]-=(qr[i][j]+qz[i][j])*dt/
                    (rhd[i][j]*cp[i][j]*dslong[i]*dz);

            temp[i+1][j][1]+=qr[i][j]*dt/
                    (rhd[i+1][j]*cp[i+1][j]*dslong[i+1]*dz);

            temp[i][j+1][1]+=qz[i][j]*dt/
                    (rhd[i][j+1]*cp[i][j+1]*dslong[i]*dz);
            }
for(i=0;i<axe_nodes;i++){
            temp[rad_nodes-1][i][0]=temp[rad_nodes-2][i][0];
            temp[rad_nodes-1][i][1]=temp[rad_nodes-2][i][1];
            }

/* Computation of surface losses */
for(i=0;i<rad_nodes;i++){
            red_loss=dslong[i]*
            (5.67e-12)*(Sqr4(temp[i][0][1]+273.0)- Sqr4(273.0));
            conv_loss=dslong[i]*(1.0e-3)*temp[i][0][1];
            temp[i][0][1]-=(red_loss+conv_loss)
            *dt/(rhd[i][0]*cp[i][0] *dslong[i]*dz);
            }
/* Transfer of the current temperature distribution
        into the matrix of the next iteration   */
for(i=0;i<rad_nodes;i++)
for(j=0;j<axe_nodes;j++){
            temp[i][j][0]=temp[i][j][1];
            }
/* printout of  results  on the PC screen
    and compute next time for printout,
    to get same values as for Figure 2.7 use "timlc=timlc+0.01"
    instead */
if(timl>=timlc){
            timlc=timlc+0.1;
```

```
            printf("\n\n Time is: %f sec,  Pass # %d\n",tim, iter);
            for(j=0;j<rad_nodes;j++){
            for(i=0;i<rad_nodes-10;i++){
                    printf("%3.3f  ",temp[i][j][1]);
                    }
                    printf("\n");
                    }

            /*printf("\n\n radial flux \n ");
            for(j=0;j<rad_nodes;j++){
            for(i=0;i<rad_nodes;i++){
                    printf(" %g  ",qr[i][j]*10.0/dslat[i]);
                    }
                    printf("\n");
                    }
            printf("\n\n axial flux\n ");
            for(j=0;j<rad_nodes;j++){
            for(i=0;i<rad_nodes;i++){
                    printf(" %g  ",qz[i][j]*10.0/dslong[i]);
                    }
                    printf("\n");
                    }*/
            }
}/*End of the iterations */

printf("\nconductivity of the %d subsurface
elements\n",def_nod_pos);
for(i=0;i<rad_nodes;i++) printf("%f ",kint[i]);

/* deallocate memory*/
free(dslong);free(kint);free(cp);free(dslat);
free(rhd);free(qr);free(qz);free(temp);return(0);
}
```

***

Appendix B
Smoothing Routine

In this appendix, we give the routines necessary to smooth data. The method is discussed in Sect. 3.2.2. The code is written in "C" language.

We give below an example of a program which can be used to smooth an image; it smooths rows and columns separately (note that only functions for the smoothing are given in module SMOOTH, C, below).

```
#define GAUSS MAXCOL /* for the sliding Gaussian */

main()
{

/*--------------- VARIABLES OF THE PROGRAM --------------------*/

/* include here the list of all variables  used by the program  */

printf("Program to smooth images");
input_parameters(); /* ask the user to input all the parameters */
 /* read the image to smooth from disk */
read_disk(matresul, filename);

  /* Processing on ROWS */
    init_gauss(y_gauss, &gauss_dim, Rlig);
    a_zero(lisse); /* set elements of vector "lisse" to zero */
    for (row = 0; row <= nb_rows; row++)
      {
      get_row(tempo, matresul, row); /* extract 1 row from image */
      lissage_gauss(lisse, tempo, dimlig, gauss_dim, y_gauss,
      nb_passe);
      save_lig(lisse, matresul, row); /* save smoothed row in the
      image */
      } /* for */
  /* Processing on COLUMNS */
{
a_zero(lisse); /* set elements of vector "lisse" to zero */
init_gauss(y_gauss, &gauss_dim, Rcol); /* another parameter for
the Gaussian */
    for (column = 0; column < MAXCOL; column++)
      {
      get_column(tempo, matresul, column, nb_lignes);
      lissage_gauss(lisse, tempo, dimcol, gauss_dim, y_gauss,
      cnb_passe);
```

```
        save_col(lisse, matresul, column);
     } /* for */
} /* if */
 /* Save the smoothed image */
save_disk(matresul, filename);
printf("\n---------- End of the smoothing program ----------\n");
} /* end of the program */


/************************************************************
**
*
*                    S M O O T H . C
*
*   Module for smoothing raw data based on a sliding Gaussian.
*   function "lissage_gauss()" which performs the smoothing on
*   a vector having a maximum of MAXCOL elements.
*   The main program should call "lissage_gauss()" routine for
*   smoothing one row/column of the image at a time.
*   All parameters are passed to the routine. Prior to the call,
*   the Gaussian is first initialized  with function "init_gauss"
*   so that the summation of all the elements is one. A uniform
*   (all elements equal) vector will come out intact of the
*   smoothing process.
*
*   Xavier Maldague + Gilles Filion
*
*************************************************************/

#include "c:\archives\disdef.h"
#define GAUSS MAXCOL   /* number of elements for the Gaussian */

/* --- functions available in this module --- */
void init_gauss(double [GAUSS], int *, double);
void lissage_gauss(double [MAXCOL], double [MAXCOL], int, int,
                   double [GAUSS], int);

/* ------------------------------------------------------------- */

void init_gauss(y_gauss, pgauss_dim, R)
double y_gauss [GAUSS];      /* vector containing Gaussian Bell */
int    *pgauss_dim;          /* nb elements of Gaussian          */
double R;                    /* parameter for Gaussian width     */
/*
** Function which computes a Gaussian and save it in a vector.
** Parameter C is adjusted to insure the summation of all elements
** is one despite the small and discrete number of points.
*/
{
/* initialization of vectors and parameters for the      **
   smoothing by sliding Gaussian                          */
double total, i_maxfl;
int i, j, i_max;
double C = 0.39894228; /* (2pi)^-0.5 */
double square;

/* initialization of the vector containing the Gaussian */
for (i = 0; i < GAUSS; i++) y_gauss [i] = 0.0;

i_maxfl = 7.0 *  R + 1.0; /* in real to be more accurate */
/* +0.5 round truncation due to the integer number of points */
i_max  = (int) i_maxfl + 0.5;

/* Determine "C" to insure summation of all elements is one */
total = 0.0;
```

```
for (i = 0; i < i_max; ++i)
    {
    square = (double) (i_maxf1/2 - i) / R;
    total += exp(-.5 * square * square);
    }
C = R / total;
/* printf("The parameter C equals %lf \n", C); */

for (i = 0; i < i_max; ++i)
    {
    square = (double) (i_maxf1/2 - i) / R;
    y_gauss[i] = C * exp(-.5 * square * square) / R;
    }
*pgauss_dim = i_max;

total = 0.0;
for (i = 0; i < i_max; ++i)
    total += y_gauss[i];
/* printf("Summation all elements of Gaussian %lf\n", total); */

} /* init_gauss */

/* ------------------------------------------------------------ */

void lissage_gauss(lisse, brut, dimension, gauss_dim, y_gauss,
nb_passe)
double  lisse [MAXCOL], brut [MAXCOL];/* vector raw + smoothed */
int     dimension; /* nb of elements to consider in the vector */
int     gauss_dim; /* nb elements to consider in the Gaussian  */
double  y_gauss [GAUSS];       /* Gaussian */
int     nb_passe;
{
double H, sortie, rebus [MAXCOL];
int i, j, index, passe;

for (passe = 1; passe <= nb_passe; passe++)
{
for (i = 0; i < dimension; i++)
    {
    H = lisse[i];
    for (j = 0; j < gauss_dim; j++)
        {
        index = i - gauss_dim / 2+j;
        if (index < 0)
            H += (brut[0] - lisse[0]) * y_gauss[j];
        else
            if (index >= dimension)
                H += (brut[dimension-1] - lisse[dimension-1])
                    * y_gauss[j];
            else
                H += (brut[index] - lisse[index])*y_gauss[j];
        } /* inner for */
    rebus[i] = H;
    } /* outer for */
for (i = 0; i < dimension; ++i)
    lisse[i] = rebus[i];
} /* for-passe */

} /* lissage_gauss */

/* ------------------------------------------------------------ */
```

* * *

Appendix C

Parabola Computation

The problem we want to solve is the fitting of a parabola to a data set (x_i, y_i). The parabola is given by the following function:

$$y = a + bx + cx^2 \tag{C.1}$$

Bevington (1969, p 137, Eq. 8.6) describes a least squares method for fitting the quadratic function of the N data points; he demonstrates that

$$a = \frac{1}{\Delta} \begin{vmatrix} \sum y_i & \sum x_i & \sum x_i^2 \\ \sum x_i y_i & \sum x_i^2 & \sum x_i^3 \\ \sum x_i^2 y_i & \sum x_i^3 & \sum x_i^4 \end{vmatrix}$$

$$b = \frac{1}{\Delta} \begin{vmatrix} N & \sum y_i & \sum x_i^2 \\ \sum x_i & \sum x_i y_i & \sum x_i^3 \\ \sum x_i^2 & \sum x_i^2 y_i & \sum x_i^4 \end{vmatrix}$$

$$c = \frac{1}{\Delta} \begin{vmatrix} N & \sum x_i & \sum y_i \\ \sum x_i & \sum x_i^2 & \sum x_i y_i \\ \sum x_i^2 & \sum x_i^3 & \sum x_i^2 y_i \end{vmatrix} \tag{C.2}$$

$$\Delta = \begin{vmatrix} N & \sum x_i & \sum x_i^2 \\ \sum x_i & \sum x_i^2 & \sum x_i^3 \\ \sum x_i^2 & \sum x_i^3 & \sum x_i^4 \end{vmatrix}$$

Note that this formula is valid only if uncertainties are the same through the data set; otherwise, individual standard deviations must be considered (Bevington 1969, p 137, Eq. 8.5).

Appendix D

Higher Order Gradient Computation Based on the Roberts Gradient

In this appendix, we derive a higher order gradient approximation with respect to Roberts gradient (Eq. (6.51)). We will first consider the one-dimensional case of a discrete function $f(x)$ for which we want to compute the first derivative $f'(x)$. Using a Taylor series:

$$f(x-h) = f(x) - hf'(x) + h^2/2f''(x) - h^3/6f'''(x) + h^4/24f''''(x) - \cdots$$
$$f(x+h) = f(x) + hf'(x) + h^2/2f''(x) + h^3/6f'''(x) + h^4/24f''''(x) + \cdots$$
$$f(x+2h) = f(x) + 2hf'(x) + 4h^2/2f''(x) + 8h^3/6f'''(x) + 16h^4/24f''''(x) + \cdots$$
$$f(x-2h) = f(x) - 2hf'(x) + 4h^2/2f''(x) - 8h^3/6f'''(x) + 16h^4/24f''''(x) - \cdots$$

If we multiply each line by a, b, c, and d respectively and we compute the summation, we obtain (limiting ourselves to fourth order derivatives):

$$af(x-h) + bf(x+h) + cf(x+2h) + df(x-2h) = +(a+b+c+d)f(x)$$
$$+ [(b-a) + 2(c-d)]hf' + [(b+a) + 4(c+d)]h^2/2f''$$
$$+ [(b-a) + 8(c-d)]h^3/6f''' + [(b+a) + 16(c+d)]h^4/24f'''' \qquad \text{(D.1)}$$

Since we want $f'(x)$, we set to zero the multiplicative terms of $f(x), f''(x), f'''(x)$ and $f''''(x)$ in Eq. (D.1); this leads to the following equations:

$$a+b+c+d = 0$$
$$a+b+4c+4d = 0$$
$$-a+b+8c-8d = 0$$
$$a+b+16c+16d = 0$$

which solves for $a = 8$, $b = -8$, $c = 1$, $d = -1$. If we substitute these values in Eq. (D.1), we find

$$[8f(x-h) - 8f(x+h) + f(x+2h) - f(x-2h)]$$
$$= hf'(x)[-8-8+2+2] = -12hf'(x)$$

and thus

$$f'(x) = \tfrac{1}{12}h[-f(x-2h) + 8f(x-h) - 8f(x-h) + f(x+2h)] \qquad \text{(D.2)}$$

If we take one pixel step ($h = 1$), then neglecting the scale factor:

$$f'(x) = f(x-2) - 8f(x-1) + 8f(x+1) - f(x+2) \tag{D.3}$$

In order to compute the gradient image $|G|$, we apply Eq. (D.3) in a cross fashion on both image f rows and columns as shown in Fig. 6.12b:

For $\forall i, i = 2,3,\dots,$ Maxcol-3

and for $\forall j, j = 2,3,\dots,$ Maxrow-3, compute

$$|G| = \left\{ \begin{array}{l} [f(i,j-2) - 8f(i,j-1) + 8f(i,j+1) - f(i,j+2)]^2 \\ + [f(i-2,j) - 8f(i-1,j) + 8f(i+1,j) - f(i+2,j)]^2 \end{array} \right\}^{1/2} \tag{D.4}$$

References

Note: SPIE: Society of Photo-Interpretive Engineers, Bellingham, WA

Abel IR (1977) Radiometric accuracy in a forward looking infrared system. Opt Engng 16(3): 241–248

Abutaleb AS (1989) Automatic thresholding of gray-level pictures using two-dimensional entropy. Comput Vision Graph Image Process 47(1): 22–32

Agam U, Gal E, Markevich N, Grimberg E (1988) Hot spot detection probability dependence on thermal imager parameters. In: Spiro IJ (ed) Infrared technology XIV. Proc SPIE vol 972, pp 188–194

AGEMA (1984) AGEMA thermovision 782 operating manual

Aguilera R (1987) 256 × 256 hybrid Schottky focal plane arrays. In: Buser RG, Warren FB (eds) Infrared sensors and sensor fusion. Proc SPIE vol 782, pp 108–113

Ahmed T, Feng ZJ, Kuo PK, Hartikainen J, Jaarinen J, (1987) Characterization of plasma sprayed coatings using thermal wave infrared video imaging. J Nondestruct Eval 6(4): 169–175

Ahmed T, Kuo PK, Favro LD, Jin HJ, Thomas RL, Dickie RA (1989) Parallel thermal wave IR video imaging of polymer coatings and adhesive bonds. In: Thompson DO, Chimenti DE (eds) Review of progress in quantitative non destructive evaluation vol 8A. Plenum Press, New York, pp 1385–1392

Allport J, McHugh J (1988) Quantitative evaluation of transient video thermography. In: Thompson DO, Chimenti DE (eds) Review of progress in quantitative non destructive evaluation vol 7A. Plenum Press, New York, pp 253–262

Anthony S, Ramamoorthy PA, Grogan TA (1987) Median-filters – optical implementation using symbolic substitution. In: Tescher AG (ed) Applications of digital image processing. Proc SPIE vol 829, pp 140–143

Arconada A, Argiriou A, Papini F, Pasquetti R (1987) La mesure en thermographie infrarouge: calibration et traitement du signal. J Mod Optics 34(10): 1327–1335 (in French)

Arpaci VS (1966) Conduction heat transfer. Addison-Wesley, Reading, MA

Auric D, Hanonge E, Kerrand E, de Miscault J-C (1987) Thermal imaging system for material processing. In: Kreutz E, Quenzer A, Schuocker D (eds) High power lasers. Proc SPIE vol 801, pp 354–358

Bahraman A, Chen C-H, Genecko JM, Shelstad MH, Ting RN, Vodicka JG (1987) Current state of the art in InSb infrared staring imaging devices. In: Caswell RS (ed) Infrared Systems and Components. Proc SPIE vol 750, pp 27–31

Bailey SJ (1988) Instruments connected to plant networks control process temperatures. Control Eng (Jan): 57–61

Baker DC, Aggarwal JK, Hwang SS (1988) Geometry guided segmentation of outdoor scenes. In: Trivedi MM (ed) Applications of artificial intelligence VI. Proc SPIE vol 937, pp 576–583

Baker IM, Crimes G, Ard C, Jenner MD, Pearsons JE, Ballingall RA, Elliott CT (1990) Photovoltaic CdHgTe – silicon focal planes. In: Proc 4th international conference on advanced infrared detectors and systems, London, June

Balageas D (1991) Le contrôle non destructif par méthodes thermiques. Rev Gén Therm Fr 300: 483–498 (in French)

Balageas DL, Luc AM (1986) Transient thermal behavior of directional reinforced composites: applicability limits of homogeneous property model. AIAA J 24(1): 109–114

Balageas DL, Krapez J-C, Cielo P (1986) Pulsed photothermal modeling of layered materials. J Appl Phys 59(2): 348–357

Balageas DL, Déom AA, Boscher D (1987a) Contrôle nondestructif des composites carbone-époxy par méthode photothermique impulsionnelle. Rev Gén Therm Fr 301: 37–41 (in French)

Balageas DL, Bosher DM, Déom AA (1987b) Temporal moment in pulsed photothermal radiometry: application to carbon epoxy NDT. In: 5th international topical meeting on photoacoustic and photo-thermal phenomena, Heidelberg RFA, 27–30 July. Springer, Berlin Heidelberg New York, pp 500–502 (Springer series in optical sciences, vol 58)

Balageas DL, Déom AA, Bosher DM (1987c) Characterization and nondestructive testing of carbon-epoxy composites by a pulse photothermal method. Mater Eval 45(4): 461–465

Balageas DL, Bosher D, Déom A, Fournier J, Henry R (1988) La thermographie infrarouge: un outil quantitatif à la disposition du thermicien. Rev Gén Therm Fr 322: 501–510 (in French)

Balageas DL, Bosher DM, Déom AA (1990) Measurement of convective heat-transfer coefficients on a wind tunnel model by passive and stimulated infrared thermography. Infrared technology XVI. Proc SPIE vol 1341

Balageas DL, Bosher DM, Déom AA (1991) Photoacoustic microscopy by photodeformation applied to the determination of thermal diffusivity. In: Baird GS (ed) Thermosense XIII. Proc SPIE vol. 1467, pp 278–289

Ballingall RA (1990) Review of infrared focal plane arrays. In: Lettington AH (ed) Infrared technology and applications. Proc SPIE vol 1320, pp 70–87

Bantel T, Bowman D, Halase J, Kenve S, Krisher R, Sippel T (1986) Automated infrared inspection of jet engine turbine blades. In: Kaplan H (ed) Thermosense VIII. Proc SPIE vol 581, pp 18–23

Barnard KJ, Boreman GD (1990) Synthesis of infrared spectral signatures. Opt Engng 29(3): 233–239

Barnsley MF, Sloan AD (1988) A better way to compress images. Byte (Jan): 215–223

Barre JC (1988) Thermographie infrarouge et maintenance. Achats et entretien 405 (Mar): 42–48 (in French)

Baughn TV, Johnson DB (1986) A method for quantitative characterization of flaws in sheets by use of thermal-response data. Mater Eval 44: 850–858

Beaudoin JL, Bissieux C (1992) Theoretical aspects of the infrared radiation. In: Maldague XPV (ed) Infrared methodology and technology, chap 2. Gordon and Breach, New York (International advances in NDT monograph series) (in press)

Beaudoin JL, Merienne E, Dartois R, Danjoux R (1985) Quantitative photothermal imaging: A new method for the non-invasive characterization of thermally thin and layered materials. In: Proc fifth infrared information exchange, New Orleans, 29–31 October, pp 11–18

Bedrossian J, Slazas P (1987) Practical applications of infrared thermometry. Sensors (Jan): 24–31

Beghdadi A, Le Negrate A (1989) Contrast enhancement techniques based on local detection of edges. Comput Vision Graph Image Process 46(2): 162–174

Bell IG (1991) A high performance infrared thermography system based on class II thermal imaging common modules. In: Baird GS (ed) Thermosense XIII. Proc SPIE vol 1467, pp 438–447

Bell JF, Hamilton-Brown R, Hawke BR (1988) Composition and size of Apollo asteroid 1984 KB. Icarus 73(3): 482–486

Bell ZW (1987) An image segmentation algorithm for non film radiography. In: Thompson DO, Chimenti DE (eds) Review of progress in quantitative non destructive evaluation, vol 6A. Plenum Press, New York, pp 773–778

Bell ZW (1988) Evaluation of tresholding heuristics useful for automated filmless radiography. In: Thompson DO, Chimenti DE (eds) Review of progress in quantitative non destructive evaluation, vol 7A. Plenum Press, New York, pp 739–746

Beniger JR, Robyn DL (1978) Quantitative graphics in statistics: a brief history. Am Statist 32(1): 1–11

Bentz CP, Martin JW (1992) Thermographic imaging of surface finish defects in coatings on metal substrates. Mater Eval 50(2): 242–246

Bevington PR (1969) Data reduction and error analysis for the physical sciences. McGraw Hill, New York

Beynon TGR (1982) Radiation thermometry applied to the development and control of gas turbine engines. In: Schooley JF (ed) Temperature, its measurement and control in science and industry, vol 5 part 1. American Institute of Physics, pp 471–477

Bieman LH (1988) Three-dimensional machine vision. Photon Spect (May): 81–85.

Biermann R (1988) Thermocouples. Stand News 16(5): 40–42

Blais F, Rioux M, Beraldin JA (1988) Practical considerations for a design of a high precision 3-D laser scanner system. In: Optomechanical and electro-optical design of industrial systems. Proc SPIE vol 959, pp 225–246

Boivin D, Laurendeau D, Comeau F, Richards C (1989) A computer-vision based apparatus for the measurement of planar movement: an application in physiotherapy. In: Proc Vision Interface '89, pp 37–44

Boogaard J (1992) Need and necessity of NDT. In: Maldague XPV (ed) Infrared methodology and technology, chap. 1. Gordon and Breach, New York (International advances in NDT monograph series) (in press)

Bosher DM, Balageas DL, Déom AA, Gardette G. (1988) Non destructive evaluation of carbon epoxy laminates using transient infrared thermography. In: Proc 19th symposium on nondestructive evaluation, NTIAC, San Antonio, TX

Bouchardy AM, Durand G, Gauffre G (1983) Processing of infrared thermal images for aerodynamic research. Report ONÉRA, tp 1983-32, SPIE, Geneva, 18–22 April 1983

Bow ST (1984) Pattern recognition. Marcel Dekker, New York

Bracewell RN (1984) The fast Hartley transform. Proc IEEE 72(8): 1010–1018

Braim SP, Cuthbertson GM (1986) A dual waveband imaging radiometer. In: Infrared technology XII. Proc SPIE vol 685, pp 129–137

Brownson J, Gronokowski K, Meade E (1987) Two-color imaging radiometry for pyrotechnic diagnostics. In: Madding RP (ed) Thermosense IX. Proc SPIE vol 780, pp 194–201

Bruckstein AM (1988) On shape from shading. Comput Vision Graph Image Process 44(2): 139–154

Buchanan RA, Condon P, Klynn L (1990) Recent advances in digital thermography for nondestructive evaluation. In: Semanovich SA (ed) Thermosense XII. Proc SPIE vol 1313, pp 134–142

Bumbaca F, Smith KC (1988) A practical approach to image registration for computer vision. Comput Vision Graph Image Process 42: 220–233

Burch SF (1987) Digital enhancement of video images for NDT. NDT Int 20(1): 51–56

Burch SF, Burton JT, Cocking SJ (1984) Detection of defects by transient thermography: a comparison of predictions from two computer codes with experimental results. Br J NDT (Jan): 36–39

Burger C, Babak R (1985) Nondestructive evaluation through transient thermographic imaging. In: 15th symposium on NDE, San Antonio, TX, 23–25 April, pp 56–67

Burgess DE, Manning PA, Watton R (1985) The theoretical and experimental performance of a pyroelectric array imager. Infrared technology XI. Proc SPIE vol 572, pp 2–6

Burleigh D (1987) A bibliography of NDT of composite materials performed with infrared thermography and liquid crystals. Thermosense IX. Proc SPIE vol 780, pp 250–255

Burleigh D (1988a) Bibliography of the application of infrared thermography to electronic and micro-electronic circuits. In: Thermosense X. Proc SPIE vol 934, pp 99–100

Burleigh D (1988b) Bibliography of the application of infrared thermography to welding. In: Thermosense X. Proc SPIE vol 934, pp 190–193

Burleigh D, De La Torre W (1991) Thermographic analysis of anisotropy in the thermal conductivity of composite materials. In: Baird GS (ed) Thermosense XIII. Proc SPIE vol 1467, pp 303–310

Burton M, Benning C (1981) Comparison of imaging infrared detection algorithms. In: Narendra PM (ed) Infrared technology for target detection and classification. Proc SPIE vol 302, pp 26–32

Calais E, House WR (1990) IR imaging cuts industry losses. Photon Spectra 29 (6): 87–96

Carlslaw HS, Jaeger JC (1959) Conduction of heat in solids. 2nd ed. Oxford University Press

Cashdollar KL, Hertzberg M (1982) Infrared temperature measurements of gas and dust explosion. In: Schooley JF (ed) Temperature: its measurement and control in Science and Industry, vol 5, part 1. American Institute of Physics, pp 453–462

Chambers JB (1984) Thermographic evaluation of bond lines and materials consistency of composites. In: Thermosense VII. Proc SPIE vol 520, pp 92–95

Charbonnier F, Lepoutre F, Roger JP, Lemoine A, Robert P (1989) The 'mirage' sensor in a industrial environment: optical and thermal losses determinations. In: Thompson DO, Chimenti DE (eds) Review of progress in quantitative nondestructive evaluation, vol 8A. Plenum Press, New York, pp 1105–1110

Chen CH (ed) (1988) Signal processing handbook. Marcel Dekker, New York

Chen JS, Medioni G (1989) Detection, localization, and estimation of edges. IEEE Trans Patt Anal Mach Intell 11(2): 191–198

Chen L, Yang BT, Hu X-T (1990) Design principle for simultaneous emissivity and temperature measurements. Opt Engng 29(12): 1445–1448

Chow CK, Kaneko T (1972) Automatic boundary detection of the left ventricule from cineorgiograms. Comput Biomed Res 5: 388–410

Church EL, Zavada JM (1976) J Opt Soc Am 66: 1136A

Cielo P (1983) Analysis of pulsed thermal inspection. In: Moore DW, Matzkanin GA (eds) Proc 14th symposium on nondestructive evaluation, NTIAC, San Antonio, TX, 19–21 April

Cielo P (1984) Pulsed photothermal evaluation of layered materials. J Appl Phys 56(1): 230–234

Cielo P (1988) Optical techniques for industrial inspection. Academic Press, San Diego

Cielo P (1989) NDE of industrial materials. CIM Bull 82(928): 81–89

Cielo P (1992) Temperature sensing of the variable emissivity in steel sheets during annealing using a novel cavity measurement technique. In: Eklund JK (ed) Thermosense XIV. Proc SPIE vol 1682, pp 142–154

Cielo P, Maldague X, Rousset G, Jen CK (1985a) Thermoelastic inspection of layered materials: dynamic analysis. Mater Eval 43(9): 1111–1116

Cielo P, Lewak R, Balageas DL (1985b) Thermal sensing for industrial quality control. In: Kaplan H (ed) Thermosense VIII. Proc SPIE vol 581, pp 47–54

Cielo P, Jen CK, Maldague X (1986a) The converging-surface-acoustic-wave technique: analysis and applications to nondestructive evaluation. Can J Phys 64(9): 1324–1329

Cielo P, Lewak R, Maldague X (1986b) Thermal methods of NDE. Can Soc Non Destruct Test J 7(2): 30–49

Cielo P, Maldague X, Johar S, Lauzon B (1986c) Some laser-based techniques for the characterization of sintered ceramics. Mater Eval 44(6): 770–774

Cielo P, Maldague X, Déom AA, Lewak R (1987a) Thermographic nondestructive evaluation of industrial materials and structures. Mater Eval 45(6): 452–460

Cielo P, Maldague X, Krapez J-C, Lewak R (1987b) Optics-based techniques for the characterization of composites and ceramics. In: Bussiére JF, Monchalin JP, Ruud CO, Green RE (eds) Nondestructive characterization of materials II, Plenum, New York, pp 733–744

Cielo P, Krapez J-C, Lamontagne M (1988a) Lumber moisture evaluation by a reflective cavity photo-thermal technique. Revue de physique appliquée 23: 1565–1576.

Cielo P, Krapez J-C, Maldague X (1988b) Enhanced thermographic imaging for subsurface flaw detection by full-field heating of the inspected surface. In: Hess P, Pelzl J (eds) Photoacoustic and photothermal phenomena, Springer, Berlin Heidelberg New York, pp 404–407 (Springer series in optical sciences, vol 58)

Cielo P, Maldague X, Krapez J-C (1991) Device for subsurface flaw detection in reflective materials by thermal-transfer imaging. American patent 4 996 426 (filed 26 Feb)

Coates PB (1981) Multiwavelength pyrometry. Metrologia 17: 103–109

Coester JY (1988) Technologies infrarouges. Onde Élect 68(2): 40–44 (in French)

Cohen E (1973) The design and application of the traversing infra-red inspection system–TIRIS. Non-destruct Test (Apr): 74–80

Coster M, Chermant JL (1989) Précis d'analyse d'images. Presses du CNRS (in French)

Crabol P (1987) Intérêt des polymères pyroélectriques. Contrôle Indust Qual 26 (143) (in French)

Craig DM, Chapman CE (1991) NDI of impact damaged composite panels. Br J NDT 33(2): 64–68

Czerny M (1929) Z Phys 53(1) (in German)

Dance WE, Middlebrook JB (1978) Neutron radiographic nondestructive inspection for bonded composite structures. Nondestructive evaluation and flaw criticality for composite materials. ASTM publication vol 696, pp 57–71

Danjoux R, van Schel E, Potier F, Beaudoin JL, Égee M (1987) Caractérisation x-y-z de matériaux minces par thermographie et sonde photothermique à balayage. Contrôle Indust Qual 26(147): 54–58 (in French)

Dartois R, Égee M, Marx J, van Schel E (1987) Caractérisation des matériaux bi-couches semi-transparents par radiométrie photothermique. Rev Gén Therm Fr 301: 22–32 (in French)

David D, Marin JY, Tretout H (1992) Automatic defects recognition in composite aerospace structures from experimental and theoretical analysis as part of an intelligent infrared thermographic inspection system. In: Eklund JK (ed) Thermosense XIV, SPIE vol 1682, pp 182–193

Degiovanni A (1986) Une nouvelle technique d'identification de la diffusivité thermique pour la méthode « flash ». Rev Phys Appl 21: 229–237 (in French)

Degiovanni A (1988) Conduction dans un "mur" multicouche avec sources: extension de la notion de quadripole. Int J Heat Mass Transfer 31(3): 553–557 (in French)

Délouard P, Marin J-Y, Avenas-Payan I, Tretout H (1989) Infrared thermography development for composite material evaluation. In: Boogaard J Van Dijk GM (eds) Non-destructive testing, Elsevier, Proc 12th world conference on nondestructive testing, Amsterdam, 23–28 April, pp 562–572

Delpech PM, Balageas DL (1991) Mesure par thermographie infrarouge stimulée de résistances thermiques d'interface dans des structures bonnes conductrices de la chaleur. Report of the Société Française des thermiciens, 9th January (in French)

Delpech Ph, Bosher D, Déom A, Balageas D (1990) Utilisation de la transformation de Laplace pour la détermination des grandeurs thermiques. Colloque de la Société Française des Thermiciens, ONERA TP-42 (in French)

Deriche R (1990) Fast algorithms for low-level vision. IEEE Trans Patt Anal Mach Intell 12(1): 78–87

DeWitt DP (1986) Inferring temperature from optical radiation measurements. Opt Engng 25(4): 569–601

Dickstein PA, Spelt JK, Sinclair AN, Bushlin Y (1991) Investigation of nondestructive monitoring of the environmental degradation of structural adhesive joints. Mater Eval 49(12): 1498–1499

Ding K (1985) Test of jet turbine blades by thermography. Opt Engng 24(6): 1055–1059

Dixon J (1988) Radiation thermometry. J Phys E Sci Instrum 21: 425–436

Doering ER, Basart JP (1988) Trend removal in x-ray images. In: Thompson DO, Chimenti DE (eds) Review of progress in quantitative non destructive evaluation, vol 7A. Plenum Press, New York, pp 785–794

Dorey G (1988) Effects of defects on advanced composite performance. Met Mater 4(5): 286–289

Dresser D (1990) IR Imaging for printed circuit boards (PCB) Testing. Imag April 5(4): pp 46–48

Dubrovskii AV, Kober VI, Mnatsakanyan EA, et al (1990) System for automatic experimental investigations of active thermovision inspection of materials and components. Sov J NDT 25(8): 611–614

Duda RO, Nitzan D, Barrett P (1979) Use of range and reflectance data to find planar surface regions. IEEE Trans Patt Anal Mach Intell 1(3): 259–271

Dufour M, Maldague X (1987) Prevention of spatter and molten particles emission on protective windows in welding applications. Weld J 66(6): 43–46

Dumpert DT (1992) Infrared techniques for printed circuit boards (PCB) evaluation. In: Maldague XPV (ed) Infrared methodology and technology, chap 7. Gordon and Breach, New York (International Advances in NDT Monograph Series) (in Press)

Durrani TS, Rauf A, Boyle K, Lotti F, Baronti S (1987) Thermal imaging techniques for the non destructive inspection of composite materials in real time. Proc ICASSP 87 (IEEE), pp 598–601

Elliot CT (1981) New detector for thermal imaging systems, Electron Lett 17: 312–313

Elliott H, Derin H, Cristi R (1986) Applications of the Gibbs distribution to image segmentation. In: Wegman EJ, DePriest DJ (eds) Statistical image processing and graphics. Marcel Dekker, New York, pp 3–24

Eshera MA, Don HS, Matsumoto K, Fu KS (1986) A syntactic approach for SAR image analysis. In: Wegman EJ, DePriest DJ (eds) Statistical image processing and graphics. Marcel Dekker, New York, pp 71–92

Favro LD, Ahned T, Crowther D, Jin HJ, Kuo PK, Thomas RL, Wang X (1991) Infrared thermal-wave studies of coatings and composites. In: Baird GS (ed) Thermosense XIII. Proc SPIE vol 1467, pp 290–294

Feest EA (1988) Exploitation of the metal matrix composites concept. Met Mater 4(5): 273–278

Feigenbaum V (1990) Total quality: an international imperative. Photon Spect 24(8): 82–83

Fiorini AR, Fumero R, Marchesi R (1982) Cardio-surgical thermography. In: Tesher AG (ed) Applications of digital image processing IV. Proc SPIE vol 359, pp 249–256

Florin C (1987) Thermal testing methods as new tool in NDT. Thermosense IX. Proc SPIE vol 780, pp 76–83

Fort C, Roux JM, Guidon M (1988) Les aspects thermiques des matériaux: élaboration, mise en forme, propriétés d'usage. Meeting of Societé Française des thermiciens, 3–5, May (in French)

Fraedrich DS (1991) Methods in calibration and error analysis for infrared imaging radiometers. Opt Engng 30(11): 1764–1770

Fredal D, Sega RM, Norgard JD, Bussey PE (1987) Hardware and software advancement for infrared detection of microwave fields. In: Weathersby MR (ed) Infrared image processing and enhancement. Proc SPIE vol 781, pp 160–167

Freeman H (1986) Survey of image processing applications in industry. In: Cantoni V, Levialdi S, Musso G (eds) Image analysis and processing. Plenum Press, New York, pp 1–9

Frock BG (1989) Marr-Hildreth enhancement of NDE images. In: Thompson DO, Chimenti DE (eds) Review of progress in quantitative non destructive evaluation, vol 8A. Plenum Press, New York. pp 701–708

Fu KS, Mui JK (1981) A survey on image segmentation. Patt Recog 13: 3–16

Fuchs EA, Mahin KW, Ortega AR, et al. (1991) Thermal diagnosis for monitoring welding parameters in real time. In: Baird GS (ed) Thermosense XIII. Proc SPIE vol 1467, pp 136–149

Gao M, Kari MM, Zheng S (1991) Device non specific dynamic performance model for thermal imaging systems. Opt Engng 30(11): 1779–1783

Gartenberg E, Roberts JAS (1990) Influence of temperature gradients on the measurement accuracy of IR imaging systems. In: Semanovich SA (ed) Thermosense XII. Proc SPIE vol 1313, pp 218–221

Garvie AM, Sorell GC (1990) Video-based analog tomography. Rev Sci Instrum 61(1): 138–145

Gauffre G (1986) New infrared detector and image quality. Image detection and quality. Proc SPIE vol 702, pp 39–46

Gaussorgues G (1984a) La thermographie dans les procédés industriels. In: 3rd European conference on NDT, Firenza, 15–18 October, pp 377–388 (in French)

Gaussorgues G (1984b) La thermographie infrarouge. Éditions Lavoisier, Paris (in French)

Gayer A, Saya A, Shiloh A (1990) Automatic recognition of welding defects in real-time radiography. NDT Int 23(3): 192–196

Gilmore JF (1985) Artificial intelligence in image processing. Digital image processing. Proc SPIE vol 528, pp 192–201

Girard G, Algazi VR (1985) Traitement numérique d'images infrarouges. Trait Signal 2(1): 29–43 (in French)

Gitzhofer F, Martin C, Fauchais P (1987) Contrôle par thermographie infrarouge de l'apparition de fissures dans un matériau céramique projeté par plasma et soumis à un cyclage thermique. Rev Gén Therm Fr 301: 63–69 (in French)

Gonzalez RC, Wintz P (1987) Digital image processing, 2nd edn. Addison-Wesley, Reading MA

Goss AJ (1987) The pyroelectric vidicon – a review. Passive infrared systems and technology. Proc SPIE vol 807, pp 25–31

Gostelow CR, Crocker RL, Saffari N, Bond LJ (1986) Improved ultrasonic methods for gas turbine NDE. In: Thompson DO, Chimenti DE (eds) Review of progress in quantitative non destructive evaluation, vol 5A. Plenum Press, New York, pp 145–148

Green DR (1968) Principles and application of emittance-independent infrared non destructive testing. Appl Optics 7(9): 1779

Green JE (1970) Computer methods for erythrocyte analysis. IEEE conference record of the symposium on feature extraction and selection in pattern recognition, Argonne, Ill, 5–7 October, pp 100–109

Gregory VT (1982) Adaptative histogram equalization and its applications. In: Tesher AG (ed) Applications of digital image processing IV. Proc SPIE vol 359, pp 204–209

Grinzato E, Mazzodli A (1991) Infrared detection of moist areas in monumental buildings based on the thermal inertia analysis. In: Baird GS (ed) Thermosense XIII. Proc SPIE vol 1467, pp 75–82

Gubala M, Teague JR, Burns JM, Di Marco JS (1989) Field portable HgCdTe MWIR staring array imaging system. Proc SPIE vol 1050, 105–111

Haddon JF (1988) Generalised threshold selection for edge detection. Patt Recog 21(3): 195–203

Hall EL (1979) Computer image processing and recognition. Academic Press, San Diego

Harding KG (1988) Optical considerations for machine vision. Industrial Technology Institute, Ann Arbor, MI, June

Harper BM, Norman TD, Exley DR (1990) Hydrogen fire detection using thermal imaging and its application to space launch vehicles. In: Semanovich SA (ed) Thermosense XII. Proc SPIE vol 1313, pp 309–320

Hayden HC (1987) Data smoothing routine. Comput Phys 1(1): 74–75

Heath DM, Winfree WP (1989) Thermal diffusivity measurement in carbon–carbon composites. In: Thompson DO, Chimenti DE (eds) Review of progress in quantitative non destructive evaluation, vol 8B. Plenum Press, New York, pp 1613–1619

Herby G (1988) Traitements d'images infrarouge. Onde Élect 68(2): 53–57 (in French)

Herschel W (1800a) Experiments on the solar, and on the terrestrial rays that occasion heat; with a comparative view of the laws to which light and heat, or rather the rays which occasion them, are subject, in order to determine whether they are the same or different. Phil Trans R Soc London, 90: 293–437

Herschel W (1800b) Investigation of the powers of the prismatic colours to heat and illuminate objects; with remarks, that prove the different refrangibility of radiant heat. To which is added, an inquiry into the method of viewing the sun advantageously, with telescopes of large apertures and high magnifying powers. Phil Trans R Soc London, 90: 255

Hershey W, Kim JM (1990) Impulse response applications to nondestructive testing. In: McGonnagle WJ (ed) International Advances in NDT vol 15, Gordon and Breach, New York, pp 289–311

Hobbs C, Kenway-Johnson D, Milne J (1991) Quantitative measurement of thermal parameters over large areas using pulse video thermography. In: Baird GS (ed) Thermosense XIII. Proc SPIE vol 1467, pp 264–277

Holmsten D (1986) Thermographic sensing for on-line industrial control. In: Cielo P (ed) Optical techniques for industrial inspection Proc SPIE vol 665, pp 75–87

Holmsten D, Houis R (1990) High-resolution thermal scanning for hot-strip mills. In: Semanovich SA (ed) Thermosense XII. Proc SPIE vol 1313, pp 322–331

Holst GC, Pickard JW (1989) Analysis of observer minimum resolvable temperature responses. In: Huber AJ, Triplett MJ, Wolverton JR (eds) Imaging infrared: scene simulation, modeling, and real time image tracking. Proc SPIE vol 1110, pp 252–258

Horn BKP (1986) Robot vision. McGraw Hill, New York

Houlbert A-S (1991) Détection de défauts subsurfaciques dans les matériaux composites par thermographie infrarouge. Report LEMTA (Université de Nancy) no 875 (in French)

Hsieh CK, Kassab AJ (1986) A general method for the solution of inverse heat conduction problems with partially unknown system geometries. Int J Heat Mass Transfer 29(1): 47–58

Hudson B (1985) Modern techniques in non-destructive testing. Met Mater 1(2): 88–90

Hudson RD (1969) Infrared system engineering. Wiley Interscience, New York 1969

Hughett P (1987) Image processing software for real time quantitative infrared thermography. In: Madding RP (ed) Thermosense IX. Proc SPIE vol 780, pp 176–183

Hunter GB, Allemand CD, Eagar TW (1985) Multiwavelength pyrometry: an improved method. Opt Engng 24(6): 1081–1085

Hunter GB, Allemand CD, Eagar TW (1986) Prototype device for multiwavelength pyrometry. Opt Engng Nov 25(11): 1222–1231

Hurley TL (1992) Infrared techniques for electric utilities. In: Maldague XPV (ed) Infrared methodology and technology, chap 8. Gordon and Breach, New York (International advances in NDT monograph series) (in press)

Imhof RE, Birch DJS, Moksin MM, Webb J, Willson PH, Strivens TA (1991) Thermal wave NDE. Br J NDT, April 33(4): 172–176

Jacoby MH, Lingenfelter DE (1989) Monitoring the performance of industrial computed tomography inspection systems. Mater Eval 47: 1196–1199

Jain R, Martin WN, Aggarwal JK (1979) Segmentation through the detection of changes due to motion. Comput Vision Graph Image Process 11: 13–34

Jakowatz CV, Smiel AJ, Eichel PH (1987) Pyroelectric line scanner for remote IR imaging of vehicles. In:

Spiro IJ (ed) Infrared technology XIII. Proc SPIE vol 819, pp 36–41

James PH, Welch CS, Winfree WP (1989) A numerical grid generation scheme for thermal stimulation in laminated structures. In: Thompson DO, Chimenti DE (eds) Review of progress in quantitative non destructive evaluation, vol 8A. Plenum Press, New York, pp 801–809

Jarem JM, Pierluissi JH, Ng WW (1984) A transmittance model for atmospheric methane. In: Spiro IJ Mollicone RA (eds) Infrared technology X. Proc SPIE vol 510, pp 94–100

Jen CK, Cielo P, Maldague X (1985) Non-contact ultrasonic characterization of piezoelectric ceramics. J Am Ceram Soc 68(6): C-146–C-148

Johnson RB, Feng C, Fehribach JD (1988) On the validity and techniques of temperature and emissivity measurements. In: Lucier RD (ed) Thermosense X. Proc SPIE vol 934, pp 202–206

Jost SR, Meikleham VF, Myers TH (1987) InSb: a key material for IR detector applications. Mat Res Soc Symp Proc vol 90, pp 429–435

Kaplan H (1987) Process control using IR sensors and scanners. Photon Spect (Dec): 92–95

Kapur JN, Sahoo PK, Wong AKC (1985) A new method for gray-level picture tresholding using the entropy of the histogram. Comput Vision Graph Image Process 29: 273–285

Kassab AJ, Hsieh CK (1987) Application of infrared scanners and inverse heat conduction methods to infrared computerized axial tomography. Rev Sci Instrum 58(1): 89–95

Kennedy HV (1991) Modeling second-generation thermal imaging systems. Opt Engng 30(11): 1771–1778

Kiliski S (1991) Magnetic disk storage for video images. Adv Imag 6(5): 32–34

Kimata M, Denda M, Yutani N, Iwade S, Tsubouchi N (1988) High density schottky-barrier infrared image sensor. Thermosense X. Proc SPIE vol 930

Kirkwood JJ (1991) Behavioral observations in thermal imaging of the big brown bat, *Eptesicus fuscus*. In: Baird GS (ed) Thermosense XIII. Proc SPIE vol 1467, pp 369–371

Kobayashi A, Ueda S (1987) Development of television camera for gas pipeline inspection. In: Kittmer A (ed) 5th Pan Pacific conference on NDT, Vancouver, April, pp 391–402

Koehler T (1991) Infrared detectors continue to diversify. Laser Focus World 27(3): A31–A34

Kraft GD, Wing NT (1981) Microprogrammed control and reliable design of small computers. Prentice-Hall, Englewood Cliffs, NJ

Krapez J-C (1991) Contribution à la caractérisation des défauts de type délaminage ou cavité par thermo-graphie stimulée. PhD thesis, École Centrale des Arts et Manufacture de Paris, Chatenay-Malabry, France, March (in French)

Krapez J-C, Cielo P (1991) Thermographic nondestructive evaluation: data inversion procedures part I: 1D analysis. Res Nondestruct Eval 3(2): 81–100

Krapez J-C, Cielo P, Maldague X, Utracki LA (1987) Optothermal analysis of polymer composites. Polym Compos 8(2): 396–407

Krapez J-C, Cielo P, Lamontagne M (1990) Reflective-cavity infrared temperature sensors: an analysis of spherical, conical and double wedge geometries. In: Lettington AH (ed) Infrared technologies and applications. Proc SPIE vol 1320, pp 186–201

Krapez J-C, Maldague X, Cielo P (1991) Thermographic nondestructive evaluation: data inversion procedures part II: 2-D analysis and experimental results. Res Nondestruct Eval 3: 101–124

Krishnakumar N, Sitharama S, Hoylder R, Lybanon M (1990) Feature labelling in infrared oceanographic images. Image Vision Comput 8(2): 142–147

Krishnan G, Walters D (1988) Segmenting intersecting and incomplete boundaries. In: Trivedi MM (ed) Applications of artificial intelligence VI. Proc SPIE vol. 937, pp 550–556

Krummar UKP (ed) (1991) Parallel architectures and algorithms for image understanding. Academic Press, San Diego.

Kuo PK, Ahmed T, Favro LD, Jin H-J, Thomas RL (1989) Synchronous thermal wave IR video imaging for nondestructive evaluation. J. Nondestruct Eval 8(2): 97–106

Kvernes I, Espeland M, Norholm O (1988) Plasma spraying of alloys and ceramics. Scand J Metall 17: 8–16

Lang D (1988) Initiation et propagation des endommagements dans les composites stratifiés carbone-époxy. Matér Techniques (Apr–May): 17–22 (in French)

Lau SK, Almond DP, Patel M, Corbett J, Quigley MBC (1990) Analysis of transient thermal inspection. In: Lettington AH (ed) Infrared technology and applications. Proc SPIE vol 1320, pp 178–185

Laurendeau D (1982) Acquisition automatique et traitement mathématique de formes anatomiques. MSc thesis, Universite Laval (in French)

Laurendeau D, Poussart D (1985) A segmentation algorithm for extracting 3D edges from range data. Compint 85 (IEEE), pp 765–767

Laurin TC (1990) Editorial comment: no alternative to quality. Photonics Spectra 24(8): 14

Lawson RN (1956) Implications of surface temperature in the diagnosis of breast cancer Can Med Ass J 75: 309

Lee DJ, Krile TF, Mitra S. (1987) Digital registration technique for sequential fundus images. In: Tesher

AD (ed) Applications of digital image processing X. Proc SPIE vol 829, pp 293–300

Lee JS (1980) Digital image enhancement and noise filtering by use of local statistics. IEEE Trans Patt Anal Mach Intell 2(2): 165–168

Lepoutre F, Roger J-P (1987) Mesures thermiques par effet mirage. Rev Gen Therm Fr (301): 15–21 (in French)

Leszczynski KW, Shalev S (1989) A robust algorithm for contrast enhancement by local histogram modification. Image vision Comput 7(3): 205–209

Levine BF, Choi KK, Bethea CG, Malik J (1987) New 10 µm infrared detector using intersubband in resonant tunnelling GaAlAs superlattices, Appl Phys Lett 50 (Apr): 16

Levinstein H (1965) Infrared detectors. In: Kingslake R (ed) Applied optics and optical engineering, vol II, Academic Press, pp 311–347

Lewak R (1992) Infrared techniques in the nuclear power industry. In: Maldague XPV (ed) Infrared methodology and technology, chap 9, Gordon and Breach, New York (International Advances in NDT monograph series) (in press)

Liddicoat TJ, Mansi MV (1988) An infrared radiometer using uncooled pyroelectric detectors for scientific and general use. In: Hall PR and Seeley J (eds) Infrared systems–design and testing. Proc SPIE vol 916, pp 63–68

Lin H-M, Willson AN (1988) Median filters with adaptive length. IEEE Trans Circ Syst 35(6): 675–690

Lineberry M (1982) Image segmentation by edge tracing. In: Tesher AG (ed) Applications of digital image processing IV. Proc SPIE vol 359, pp 361–367

Ljungberg S-Å (1992) Infrared techniques in buildings and structures: operation and maintenance. In: Maldague XPV (ed) Infrared methodology and technology, chap 6. Gordon and Breach, New York (International advances in NDT monograph series) (in press)

Loubet D (1987) Mesures et suivi de l'endommagement des matériaux composites. Report 872-930-110, Société Aérospatiale, Paris, October (in French)

Lozano-Garciá DF, Fernández, Johannsen CJ (1991) Assessment of regional biomass-soil relationship using vegetation index. IEEE Trans Geosci Rem Sens 29(2): 331–339

Lu YJ, Hsu YH, Maldague X (1992) Vehicle classification using infrared image analysis. J Transport Engng 118 (2): 223–240.

Lucier RD (1991) So now what ? – things to do if your IR program stops producing results. In: Baird GS (ed) Thermosense XII. Proc SPIE, vol 1467, pp 59–62

Madrid A (1990) On the estimation of hardware failure rates using IR thermography. In: Semanovich SA (ed) Thermosense XII. Proc SPIE 1313, pp 30–46

Maldague X (1992) Instrumentation for the infrared. In: Maldague XPV (ed) Infrared methodology and technology chap 3. Gordon and Breach, New York (International Advances in NDT monograph series) (in press)

Maldague X, Dufour M (1989) A dual-imager and its applications for active vision robot welding, surface inspection and two-color pyrometry. Opt Engng 28(8): 872–880

Maldague X, Poussart D, Laurendeau D, April R (1986a) Tridimensional form acquisition apparatus. In: Cielo P (ed) Proc SPIE: Optical techniques for industrial inspection, vol 665. pp 200–208

Maldague X, Cielo P, Jen CK (1986b) NDT Applications of laser-generated focused acoustic waves. Mater Eval 44(9): 1120–1124

Maldague X, Cielo P, Cole K, Vaudreuil G (1987a) Detection of rolled-in surface scale in steel sheets by optical and thermal inspection techniques. In: Penny CM, Caulfield HJ (eds) Optical techniques for industrial measurement and control, ICALEO 86 Conference. Springer-Verlag, Berlin Heidelberg New York

Maldague X, Cielo P, Ashley PJ, Farahbakhsh B (1987b) Thermographic NDT of aluminium laminates. Can Soc Non Destruct Test J, July–August 8(4): 44–50

Maldague X, Krapez J-C, Cielo P, Poussart D (1988) Processing of thermal images for the detection and enhancement of subsurface flaws in composite materials. In: Chen CH (ed) Signal processing and pattern recognition in nondestructive evaluation of materials, vol F44, Springer, Berlin Heidelberg New York, pp 257–285 (NATO ASI Series).

Maldague X, Krapez J-C, Cielo P, Poussart D (1989a) Inspection of materials and structures by infrared thermography: signal processing techniques for defect enhancement and characterization. Can Soc Non Destruct Test J 10(1): 28–36

Maldague X, Krapez J-C, Cielo P, Poussart D (1989b) Infrared thermographic inspection by internal temperature perturbation techniques. In: Boogaard J, Van Dijk GM (eds) Non-destructive testing, proc 12th world conference on non-destructive testing, Amsterdam, 23–28 April. Elsevier pp 561–566

Maldague X, Cielo P, Poussart D, Craig D, Bourret R (1990a) Transient thermographic NDE of turbine blades. In: Semanovich SA (ed) Thermosense XII. Proc SPIE vol 1313, pp 161–171

Maldague X, Krapez J-C, Poussart D (1990b) Thermographic non destructive evaluation (NDE): an algorithm for automatic defect extraction in infrared images. IEEE Trans Syst Man Cybern 20(3): 722–725

Maldague X, Cielo P, Poussart D (1990) Thermographic nondestructive evaluation (NDE) of turbine blades: methods and image processing. Indust Metrol J 1(2): 139–153

Maldague X, Krapez J-C, Cielo P (1991a) Subsurface flaw detection in reflective materials by thermal-transfer imaging. Opt Engng 30(1): 117–125

Maldague X, Krapez J-C, Cielo P (1991b) Temperature recovery and contrast computations in NDE thermographic imaging systems. J Nondestruct Eval 10(1): 19–30

Maldague X, Fortin L, Picard J (1991c) Applications of tridimensional heat calibration to a thermographic NDE station. In: Baird GS (ed) Thermosense XIII. Proc SPIE vol 1467, pp 239–251

Mallick PK (1986) Fiber-reinforced composites. Marcel Dekker, New York

Mann JM, Schmerr LW, Moulder JC (1991) Neural network inversion of uniform-field eddy current data. Mater Eval 49(1): 34–39

Mansoor AK, Allemand C, Eagar TW (1991a) Noncontact temperature measurement. I. Interpolation based techniques. Rev Sci Instrum 62(2): 392–402

Mansoor AK, Allemand C, Eagar TW (1991b) Noncontact temperature measurement. II. Least squares based techniques. Rev Sci Instrum 62(2): 403–409

Mansour TM (1983) Nondestructive thickness measurement of phosphate coatings by infrared absorption. Mater Eval 41(3): 302–308

Mao Z, Strickland RN (1988) Image sequence processing for target estimation in forward-looking infrared imagery. Opt Engng 27(7): 541–549

Marr D, Hildreth E (1980) Theory of edge detection. Proc R Soc Lond B 207: 187–217

McLaughlin PV (1985) Naval air systems command. Report Air Task A310310G/0518/4F41460000, Washington DC

McLaughlin PV (1988) Defect detection and quantification in laminated composites by EATF (passive) thermography. In: Thompson DO, Chimenti DE (eds) Review of progress in quantitative non destructive evaluation, vol 7A. Plenum Press, New York, pp 1125–1132.

McLaughlin PV, McAssey EV, Koert DN, Deitrich RC (1981) NDT of composites by thermography. Proc DARPA/AFWAL of progress in quantitative NDE, pp 60–68

McNamara DK, Ahearn JS (1987) Adhesive bonding of steel for structural applications. Int Mater Rev 32(6): 292–306

Mengers P (1986) Micro-computer based digital image processing for real-time radiography. In: Thompson DO, Chimenti DE (eds) Review of progress in quantitative non destructive evaluation, vol 5A. Plenum Press, New York, pp 825–833.

Milne JM, Carter P (1986) A study into the generation of temperature gradients in materials for detecting and measuring the depth of sub-surface cracks. Report AERE R 12382, N87 22991, Harwell UK, October

Milne JM, Carter P (1988) A transient thermal method of measuring the depths of subsurface flaws in metals. Br J NDT 30(5): 333–336

Minor LG, Sklonsky J (1981) The detection and segmentation of blobs in infrared images. IEEE Trans Syst Man Cybern 11(3): 194–201

Mitra S, Nutter BS, Krile TF, Brown RH (1988) Automated method for fundus image registration and analysis. Appl Optics 27(6): 1107–1112

Monchalin JP (1986) Optical detection of ultrasound. IEEE Trans Ultrason Ferroelect Freq Control 33: 485–499

Monti R (1986) Flow visualization and digital image processing. Von Karman institute for fluid dynamics, Chaussée de Waterloo, 72, B-1640 Rhode Saint Genèse, Belgium (Lecture Series 1986–09)

Monti R, Mannara G (1985) NDT of honeycomb structures by computerized thermographic systems. Acta Astronaut 12(6): 405–414

Monti R, Mannara G (1987) The computerized thermography for NDT in aerospace applications. Proc 4th European conference on NDT, vol 2. Pergamon, pp 1266–1279

Morisseau P, Huet J, Pauton M (1987) Dimensional and geometry analysis of turbine blades by the use of computerized tomography (CT). In: Farley JM, Nichols RW (eds) Proc 4th European conference on NDT, vol 2. Pergamon, pp 1257–1265

Nagarajan S, Chin BA (1990) Infrared image analysis for on-line monitoring of arc misalignment in gas tungsten arc welding processes. Mater Eval 48: 1469–1472

Nagarajan S, Chin BA (1992) Infrared techniques for real-time weld quality control. In: Maldague XPV (ed) Infrared methodology and technology, chap 10. Gordon and Breach, New York (International Advances in NDT monograph series) (in press)

Nagarajan S, Chen WH, Groom KN, Chin BA (1988) Infrared sensors for seam tracking and penetration depth control. In: Grover CP (ed) Optical testing and metrology II. Proc SPIE vol 954, pp 568–573

Nagarajan S, Chen WH, Groom KN, Chin BA (1989) Infrared sensing for adaptive arc welding. Weld 68(11): 462s–466s

Nandhakumar N, Aggarwal JK (1988) Integrated analysis of thermal and visual images for scenes interpretation. IEEE Trans Patt Anal Mach Intell 10(4): 469–481

Narendra P.M (1981) A separable median filter for image noise smoothing. IEEE Trans Patt Anal Mach Intell 3(1): 20–29

Nazif AM, Levine MD (1984) Low level image segmentation: an expert system. IEEE Trans Patt Anal Mach Intell 6(5): 555–577

Newman WM, Sproull RF (1979) Principles of interactive computer graphics. McGraw Hill, New York

Nelson MD, Johnson JF, Lomheim TS (1991) General noise processes in hybrid infrared focal plane arrays. Opt Engng 30(11): 1682–1700

Nickolls J (1990) The design of the MasPar MP-1, A cost effective massively parallel computer. Proc IEEE Compcon pp 11–16

Nicodemus FE (1967) Radiometry. In: Kingslake R (ed) Applied optics and optical engineering, vol IV, Academic Press, pp 263–307

Norton PR (1991) Infrared image sensors. Opt Engng 30(11): 1649–1663

Nouah A, Maldague X, Robitaille F (1993) Depth correction in transient thermography inspection of nonplanar components. In: Balageas DL (ed) Proc Quantitative Infrared Thermography (QIRT 92), Eurotherm Seminar 27, Éditions Européennes Thermique et Industrie, Paris (in press)

Nutter BS, Mitra S, Krile TF (1987) Image registration for a PC-based system. In: Tesher AG (ed) Applications of digital image processing XI. Proc SPIE vol 829, pp 214–221

Oermann RJ (1987) Radiometry using thermal images – part II – technical details. Technical report ERL-0423-TR, Electronics Research Laboratory, Defence Science and Technology Organisation, Department of Defence, Salisbury, South Australia

O'Gorman L (1988) A note on histogram equalization for optimal intensity range utilization. Comput Vision Graph Image Process 41: 229–232

Öhman C (1981) Practical methods for improving thermal measurements. In: Grot RA, Wood JT (eds) Thermosense IV. Proc SPIE vol 313, pp 204–212

Oliver DW, Brown J, Cueman K, Czechowski J, Eberhard J (1986) XIM: X-ray inspection module for automatic high speed inspection of turbine blades and automated flaw detection and classification. In: Thompson DO, Chimenti DE (eds) Review of progress in quantitative non destructive evaluation, vol 5A. Plenum Press, New York, pp 817–823

Ortolano DJ (1988) Resistance thermometry. Stand News 16(5): 52–55

Pajani D (1987a) Thermographie IR: quelle longueur d'onde choisir I. Mesures, 25 May, pp 71–76 (in French)

Pajani D (1987b) Thermographie IR: quelle longueur d'onde choisir II. Mesures, 9 June, pp 77–80 (in French)

Pajani D (1987c) La thermographie infrarouge dans l'industrie du verre. Contrôle Indust Qualité (152–bis): 55–62 (in French)

Pajani D (1991) La thermographie IR sur les bains d'électrolyse. Mesures 11(634): 45–47 (in French)

Parker WJ, Jenkins RJ, Butler CP, Abbott GL (1961) Flash method of determining thermal diffusivity, heat capacity, and thermal conductivity. J Appl Phys 32(9): 1679–1684

Patel PM, Lau SK, Almond DP (1991) A review of image analysis techniques applied in transient thermographic nondestructive testing. J Nondestruct Eval 10: 31–35

Pau LF (1983) Integrated testing and algorithms for visual inspection of integrated circuits. IEEE Trans Patt Anal Mach Intell 5(6): 602–608

Peacock RG (1988) Radiation thermometry. Stand News 16(5): 31–35

Pellegrini PW (1987) A comparison of iridium silicide and platinum silicide photodiodes. In: Buser RG, Warren FB (eds) Infrared sensors and sensor fusion. Proc SPIE vol 782, pp 93–98

Pennington KS, Moorhead II RJ (eds) (1990) Image processing algorithms and techniques. Proc SPIE vol 1244

Peralta SB, Ellis SC, Christofides C, Mandelis A, Sang H, Farahbakhsh B (1991) Photopyroelectric measurement of the thermal diffusivity of recrystallized high purity aluminum. Res Nondestruct Eval 3(2): 69–80

Perl A (1987) Reasons for poor acceptance of thermography in medical community, Proc 9th conference of the IEEE engineering in medicine and biology society, Boston, MA, 13–16 November

Poussart D, Laurendeau D (1988) 3D sensing for industrial computer vision. In: Sanz JLZ (ed) Advances in machine vision, pp 122–159

Quinn MT, Hribar JR, Ruiz RL, Hawkins GF (1988) Thermographic detection of buried disbonds. In: Thompson DO, Chimenti DE (eds) Review of progress in quantitative non destructive evaluation, vol 7A. Plenum Press, New York, pp 1117–1123

Ramelot D, Ludovicy J-M, Stolz C, Fischbach J-P (1988) Capteurs industriels pour applications basses températures. Rev Gén Therm Fr (322): 517–524 (in French)

Rantala J, Hartikainen J (1991) Numerical estimation of spatial resolution of thermal NDT techniques based on flash heating. Res Nondestruct Eval 3(3): 125–139

Rapaport S (1988) The secret art of frame grabbing. Photon spect (Apr): 137–140

Ravich LE (1988) Evaluation of electronic images. Laser Focus/Electro-Optics (Jun): 145–155

Reichenbach SE, Park SK, Narayanswamy R (1991) Characterizing digital image acquisition devices. Opt Engng 30(2): 171–177

Reynolds PM (1964) A review of multicolour pyrometry for temperatures below 1500 °C. Br J Appl Phys 15: 579–589

Reynolds WN (1985) Quality control of composite materials by thermography. Met Mater 1(2): 100–102

Reynolds WN (1986) Thermographic methods applied to industrial materials. Can J Phys 64: 1150–1154

Reynolds WN (1988) Inspection of laminates and adhesive bonds by pulse-video thermography. NDT Int 21(4): 229–232

Reynolds WN, Wells GM (1984) Video-compatible thermography. Br J NDT (Jan): 40–44

Richardson CH, Schafer RW (1987) Application of mathematical morphology to FLIR images. In: Hsing TR (ed) Visual communications and image processing II. Proc SPIE vol 845, pp 249–252

Rioux M, Bechthold G, Taylor D, Duggan M (1987) Design of a large depth of view three-dimensional camera for robot vision. Opt Engng 26(12): 1245–1250

Rodriguez AA, Mitchell OR (1989) Image segmentation by background extraction refinements. In: Casasent DP (ed) Intelligent robots and computer vision VIII: algorithms and techniques. Proc SPIE vol 1192, pp 122–134

Roellig TL, Werner MW, Becklin EE (1988) Thermal emission from Saturn's ring at 380 microns. Icarus 73(3): 574–583

Roney JE (1982) Steel surface temperature measurement in industrial furnaces by compensation for reflected radiation errors. In: Schooley JF (ed) Temperature, its measurement and control in science and industry. American Institute of Physics, pp 485–489

Rosenfeld A (1969) Picture processing by computer. Academic Press, Boston, MA

Rounds EM, Sutty G (1980) Segmentation based on second order statistics. Opt Engng 19(6): 936–940

Rousset G, Lévesque D, Bertrand L, Maldague X, Cielo P (1986) Pulsed photothermoelastic quantitative evaluation of flaws in laminates. Can J Phys 64(9): 1293–1296

Ryu ZM (1991) Measurement of point spread function of thermal imager. In: Baird GS (ed) Thermosense XIII. Proc SPIE vol 1467, pp 469–474

Sadat AB (1988) Machining of graphite epoxy composite material. SAMPE Q 19(2): 1–4

Safabakhsh R (1989) Processing infrared images for high speed power line inspection. In: McIntosh GB (ed) Thermosense XI. Proc SPIE vol 1094, pp 75–82

Sahoo PK, Soltani S, Wong AKC, Chen Y (1988) A survey of thresholding techniques. Comput Vision Graph Image Process 41: 233–260

Sanmartin M-L (1988) Matériau sandwich nouveau. Matér techniques (Jun): 3–7 (in French)

Sayers CM (1984) Detectability of defects by thermal non-destructive testing. Br J NDT (Jan): 28–33

Schachter BL, Davis LS, Rosenfeld A (1979) Some experiments in image segmentation by clustering of local feature values. Patt Recog 11: 19–28

Schmalz H (1990) Infrared visualized air turbulence. In: Semanovich SA (ed) Thermosense XII. Proc SPIE vol 1313, pp 278–281

Schott JR, Biegel JD (1987) Comparison of modelled and empirical atmospheric propagation data. In: Spiro IJ, Mollicone RA (eds) Infrared technology IX. Proc SPIE vol 430, pp 35–50

Scott IG (1990) Bridging the gap. In: McGonnagle WJ (ed) International advances in NDT, vol 15. Gordon and Breach, New York, pp 1–8

Segal E, Thomas G, Rose J (1979) Hope for solving the adhesive bond nightmare? 12th symposium on NDE, San Antonio, TX, pp 269–281

Shann T, Oakley JP (1990) Novel approach to boundary finding. Image Vision Comput 8(1): 32–36

Shepard SM, Sass DT (1990) Thermal imaging of repetitive events at above-frame-rate frequencies. Opt Engng Feb. 29(2): 105–109

Shepard SM, Sass DT, Imirowicz T (1991) Enhanced temporal resolution with a scanning imaging radiometer. Opt Engng 30(11): 1716–1719

Shepherd FD, Moorey JM (1987) Design considerations for IR staring-mode cameras. In: Wight R (ed) Electro-optical imaging system integration. Proc SPIE vol 762, pp 35–50

Shushan A, Meninberg Y, Levy I, Kopeika NS (1991) Infrared image sensors. Opt Engng 30(11): 1709–1715

Siegel R, Howell JR (1972) Thermal radiation heat transfer. McGraw Hill, New York

Silk MG (1989) Weld inspection methods. Met Mater 5(4): 192–196

Smart AE (1988) Trend in optical sensors for hostile environments. In: Bieringer RJ, Harding KG (eds) Optomechanical and electro-optical design of industrial systems, Dearborn MI, 26–30 June. Proc SPIE vol 959

Smith RL (1987) Recent developments in nondestructive testing. Met Mater 3(4): 187–191

Soucy M, Laurendeau D (1990) Generating non-redundant surface representation of 3D objects using multiple range views. In: IEEE Proc ICPR, 16–21 June, Atlantic City, NJ, pp 198–200

Sparrow EM, Chess RD (1978) Radiation heat transfer, sect 2-2. McGraw Hill, New York

Spiro IJ, Schlessinger M (1989) Infrared technology fundamentals. In: Thompson BJ (ed) Optical engineering series, vol 22. Marcel Dekker, New York

Spiro IJ (ed) (1990) Selected papers on radiometry. SPIE milestone series, vol MS 14

Stansfield SA (1986) ANGY: A rule-based expert system for automatic segmentation of coronary vessels from digital subtracted angiograms. IEEE Trans Patt Anal Mach Intell 8(2): 188–199

Stein MC, Heller WG (1989) Fractal methods for flaw detection in NDE imagery. In: Thompson DO, Chimenti DE (eds) Review of progress in quantitative non destructive evaluation, vol 8A. Plenum Press, New York, pp 689–700

Stillwell PFTC (1981) Thermal imaging. J Phys E Sci Instrum 14: 1113–1118

Stovicek D (1987) Better maintenance through thermography. Power Transmiss Des, June: 48–52

Strickland RN, Gerber MR (1986) Estimation of ship profiles from a time sequence of forward-looking infrared images. Opt Engng, August 25(8): 995–1000

Svedemar B (1985) Sewage leak detection: a novel method to detect underground sewer defects. 5th infrared information exchange, New Orleans, October, pp 79–83

Tam KC (1983) Multispectral limited-angle image reconstruction. IEEE Trans Nucl Sci 30(1): 697–701

Thomas LC (1980) Fundamentals of heat transfer. Prentice-Hall, Englewood Cliffs, NJ

Tonner PD, Tosello G (1986) Computed tomography scanning for location and sizing of cavities in valve castings. Mater Eval 44(2): 203–208

Tossell DA (1987) Numerical analysis of heat input effects in thermography. J NDE 6(2): 101–107

Tossel DA (1989) The analysis of heat input effects in passive thermographic NDE. In: Thompson DO, Chimenti DE (eds) Review of progress in quantitative non destructive evaluation, vol 8A. Plenum Press, New York, pp 1763–1770

Tou JT, Gonzalez RC (1974) Pattern recognition principles. Addison-Wesley

Touloukian YS, DeWitt DP (1970) Thermal Radiative Properties. Thermophysical properties of matter, vol 7. IFI/Plenum, New York, Washington

Tower JR (1991) Staring PtSi IR cameras: more diversity, more applications. Photon Spect 25(2): 103–106

Traycoff RB (1987) Computerized infrared thermography: clinical applications and diagnostic value. Proc 9th conference IEEE engineering in medicine and biology society, Boston, 13–16 November

Traycoff RB (1992) Medical applications of infrared thermograpahy. In: Maldague XPV (ed) Infrared methodology and technology, chap 14. Gordon and Breach, New York (International advances in NDT monograph series) (in press)

Tretout H (1987) Applications industrielle de la thermographie infrarouge au contrôle non destructif de pièces en matériaux composites. Rev Gén Therm Fr (301): 47–53 (in French)

Tretout H, Marin JY (1985) Transient thermal technique for infrared non destructive testing of composite materials. In: Baker LR, Masson A (eds) Infrared technology and applications. Proc SPIE vol 590, pp 277–284

Tretout H, David D. Marin JY, de Mol R (1991) Thermally stimulated infrared thermography for composite ceramic and metallic materials inspection. Post deadline paper (unpublished), presented Thermosense XIII, SPIE Conference, 3–5 April

Trivedi MM (1990) Selected papers on digital image processing. SPIE milestone series, vol 17

Turck J (1988) Les bases de la détection infrarouge. Onde élect 68(2): 36–39 (in French)

Vacelet H (1984) Contrôles non destructif des assemblages collés. 3rd European Conference on NDT, Firenza, 15–18 October, pp 172–191 (in French)

Vavilov V (1980) Infra-red non-destructive testing of bonded structures: aspects of theory and practice. Br J NDT (Jul): 175–183

Vavilov V (1984) Effects of the size of the scanning spot and the frequency characteristics of the photoreceiver on the sensitivity of active thermal testing. Sov J NDT: 269–271 (Plenum, New York, translated from Defektoskopiya (4): 54–57)

Vavilov V (1990) Dynamic thermal tomography: perspective field of thermal NDT. In: Semanovich SA (ed) Thermosense XI. Proc SPIE vol 1313, pp 178–182

Vavilov V (1992) Infrared techniques for material analysis and nondestructive testing. In: Maldague XPV (ed) Infrared methodology and technology, chap 5. Gordon and Breach, New York (International Advances in NDT monograph series) (in press)

Vavilov V, Taylor R (1982) Theoretical and practical aspects of the thermal nondestructive testing of bonded structures. In: Sharpe RS (ed) Research techniques in NDT, vol V, Academic Press, pp 238–279

Vavilov V, Ahmed T, Jin HJ, Favro RL, Thomas LD (1990) Experimental thermal tomography of solids by using the pulsed one-side heating. Sov J NDT (12): 60–66

Vavilov V, Degiovanni A, Didierjean S, Maillet D, Sengulye AA, Houlbert A-S (1991) Thermal testing and tomography of carbon epoxy plastics. Sov J NDT: vol 16

Vavilov V, Maldague X, Picard J, Thomas RL, Favro LD (1992) Dynamic thermal tomography: New NDE technique to reconstruct inner solids structure by using multiple IR image processing. In: Thompson DO, Chimenti DE (eds) Review of progress in quantitative non destructive evaluation, vol 11. Plenum Press, New York, 425–432

Waggener W (1987) Digital recording for electro-optical imagery. Electro-optical imaging systems integration. Proc SPIE vol 762, pp 69–85

Waggoner J (1991) Photon Spect 25(2, 3)

Wallace AM (1988) Industrial applications of computer vision since 1982. IEE Proc E 135(3): 117–136

Wang Y, Chin BA (1986) On line sensing of weld penetration using infrared thermogrpahy. In: Cielo P (ed) Optical techniques for industrial inspection. Proc SPIE vol 665, pp 314–320

Warren C (1980) Digital analysis of infrared imagery. Contemporary infrared sensors and instruments. Proc SPIE vol 246, pp 144–151

Welch CS, Winfree WP, Heath DM, Cramer D (1990) Material property measurements with post-processed thermal image data. In: Semanovich SA (ed) Thermosense XII. Proc SPIE vol 1313, pp 124–133

Wendt PD, Coyle EJ, Gallagher NC (1986) Stack filters. IEEE Trans Acoust Speech Signal Process 34(4): 898–911

Weszka JS, Nagel RN, Rosenfeld A (1974) A treshold selection technique. IEEE Trans Comput (Dec): 1322–1326

Whitlock JA, Boreman GD, Brown HK, Plogstedt AE (1991) Electrical network model for SPRITE detectors. Opt Engng 30(11): 1784–1787

Williams JH, Mansouri S, Lee S (1980) One-dimensional analysis of thermal nondestructive detection of delamination and inclusion flaws. Br J NDT (May): 113–118

Williams JH, Felenchak BR, Nagem RJ (1983) Quantitative geometric characterization of two-dimensional flaws via liquid crystals thermography. Mater Eval 41(2): 190–201

Williams T (1988) Object recognition open the eyes of machine-vision systems. Comput Des (May)

Williams TL, Davidson NT (1989) Recent advances in testing of thermal imagers. In: Huber AJ, Triplett MJ, Wolverton JR (eds) Imaging infrared: scene simulation, modeling, and real time image tracking. Proc SPIE vol 1110, pp 220–231

Williams VH and Fike DK (1987) Failure analysis of raw printed circuit boards using infrared thermography. In: Thermosense IX. Proc SPIE vol 780, pp 139–147

Wilson J (1991) Thermal analysis of the bottle forming process. In: Baird GS (ed) Thermosense XIII. Proc SPIE vol 1467, pp 219–218

Winfree WP, Welch CS (1989) Thermographic detection of delaminations in laminated structures. In: Thompson DO, Chimenti DE (eds) Review of progress in quantitative non destructive evaluation, vol 8B. Plenum Press, New York, pp 1657–1662

Wise JA (1988) Liquid-in-glass thermometers. Stand News 16(5): 48–50

Wolfe R (1988) Impact of PCs on Automated test measurements. Intell Instrum Comput (May): 148–153

Wong ET-W (1982) A study of the effect of the line spread function on the performance of an infrared scanning system. Ms thesis, University of Florida

Wood JT (1982) Description of a facility for evaluating infrared imaging systems for building applications. In: Courville GE (ed) Thermosense V. Proc SPIE vol 371, pp 246–249

Wood JT, Bentz WJ, Pohle T, Hepfer K (1976) Specification of thermal imagers. Opt Engng 15(6): 531–536

Woodward R (1991) Developments in aluminum – a review. Met Mater 7(2): 96–99

Woolaway JT (1991) New sensor technology for the 3 to 5 μm imaging band. Photon Spectra 25(2): 113–119

Yanisov VV, Yanisova LK (1984) The revealability of defects in optical synthesis of thermograms from x-ray photographs of different aspects. Sov J NDT 12: 1–9 (Plenum, New York, translated from Defektoskopiya)

Yoder PR (1968) Designing a durable system. Mach Des (23 May): 190–196. Reprinted in: O'Shea DC (ed) Selected papers on optomechanical design. SPIE milestone series, vol 770, pp 17–23

Young RWL (1985) Forest fire detection and mop-up of smoldering fires: the role of thermography in the Alberta forest service. 5th infrared information exchange, New Orleans, 29–31 October, pp 93–99

Young RWL (1992) Utilization and application of infrared techniques in forest fire detection and suppression operations. In: Maldague XPV (ed) Infrared methodology and technology, chap 13. Gordon and Breach, New York (International advances in NDT monograph series) (in press)

Zahn CT (1971) Graph theoretical methods for detecting and describing gestalt clusters. IEEE Trans Comput 20(1): 68–86

Zanio K (1990) HgCdTe on Si for hybrid and monolithic FPAs. In: Dereniak EL, Sampson RE (eds) Infrared detectors and focal plane arrays. Proc SPIE vol 1308, pp 180–193

Zheng Y, Basart JP (1987) Automatic image segmentation, modeling and restoration with a rule-based expert system. In: 20th ASILOMAR conference on signal systems, and computers. (IEEE): pp 421–425

Zheng Y, Basart JP (1988) Image analysis, feature extraction, and various applied enhancement methods for NDE X-ray images. In: Thompson DO, Chimenti DE (eds) Review of progress in quantitative non destructive evaluation, vol 7A. Plenum Press, New York, pp 795–804

Zisk SH, Wittels N (1985) Camera edge response. In: Svetkoff DJ (ed) Optics, illumination and image sensing for machine vision II. Proc SPIE vol 850, pp 9–16

Zucker SW (1976) Algorithms for image segmentation. Proc NATO advanced study institute on digital image processing and analysis. Bonas, France, pp 169–183

Index